Lecture Notes in Mechanical Engineering

Lecture Notes in Mechanical Engineering (LNME) publishes the latest developments in Mechanical Engineering—quickly, informally and with high quality. Original research or contributions reported in proceedings and post-proceedings represents the core of LNME. Volumes published in LNME embrace all aspects, subfields and new challenges of mechanical engineering.

To submit a proposal or request further information, please contact the Springer Editor of your location:

Europe, USA, Africa: Leontina Di Cecco at Leontina.dicecco@springer.com
China: Ella Zhang at ella.zhang@cn.springernature.com
India, Rest of Asia, Australia, New Zealand: Swati Meherishi at swati.meherishi@springer.com

Topics in the series include:

- Engineering Design
- Machinery and Machine Elements
- Mechanical Structures and Stress Analysis
- Automotive Engineering
- Engine Technology
- Aerospace Technology and Astronautics
- Nanotechnology and Microengineering
- Control, Robotics, Mechatronics
- MEMS
- Theoretical and Applied Mechanics
- Dynamical Systems, Control
- Fluid Mechanics
- Engineering Thermodynamics, Heat and Mass Transfer
- Manufacturing Engineering and Smart Manufacturing
- Precision Engineering, Instrumentation, Measurement
- Materials Engineering
- Tribology and Surface Technology

Indexed by SCOPUS, EI Compendex, and INSPEC.

All books published in the series are evaluated by Web of Science for the Conference Proceedings Citation Index (CPCI).

To submit a proposal for a monograph, please check our Springer Tracts in Mechanical Engineering at https://link.springer.com/bookseries/11693.

Sanjay Yadav · Naveen Garg · Mukesh Kumar ·
Shankar G. Aggarwal · Shiv Kumar Jaiswal ·
Manoj Kumar
Editors

Advances in Metrology

Select Proceedings of AdMet 2024

Springer

Editors
Sanjay Yadav
Metrology Society of India (MSI)
New Delhi, Delhi, India

CSIR-NPL
New Delhi, Delhi, India

Mukesh Kumar
Faculty of Sciences and Technology
Gurugram University
Gurugram, Haryana, India

Shiv Kumar Jaiswal
CSIR-NPL
New Delhi, Delhi, India

Naveen Garg
CSIR-NPL
New Delhi, Delhi, India

Shankar G. Aggarwal
CSIR-NPL
New Delhi, Delhi, India

Manoj Kumar
CSIR-NPL
New Delhi, Delhi, India

ISSN 2195-4356 ISSN 2195-4364 (electronic)
Lecture Notes in Mechanical Engineering
ISBN 978-981-96-6417-7 ISBN 978-981-96-6418-4 (eBook)
https://doi.org/10.1007/978-981-96-6418-4

This Springer imprint is published by the registered company Springer Nature Singapore Pte Ltd.
The registered company address is: 152 Beach Road, #21-01/04 Gateway East, Singapore 189721, Singapore

Contents

About the Editors

Dr. Sanjay Yadav is the former chief scientist and head of Physico-Mechanical Metrology Division and professor-AcSIR, Faculty of Physical Sciences at CSIR-National Physical Laboratory (CSIR-NPL). Presently, he is the vice president of Metrology Society of India, New Delhi, and vice president of Ultrasonic Society of India, New Delhi, India. He has also contributed in several national and international advisory boards/committees like IMEKO, APMP, BIPM, BIS, NCERT, FCRI, etc. Prior to joining CSIR-NPL, India, he has worked for more than seven years as a project officer in the Science and Technology Department of Haryana Government for the implementation of various rural development programs. He has more than 450 research publications to his credit in national and international journals of repute, books, monographs, reports, conferences, seminars, and workshops.

Dr. Naveen Garg is a senior principal scientist and head, Physio Mechanical Metrology Division and Acoustics and Vibration Standards in CSIR-National Physical Laboratory, New Delhi, working in the field of sound and vibrations since the past 19 years. He is a mechanical engineer specializing in machine design, vibrations and acoustics, measurement science and working in development and upgradation of primary standards of sound pressure and vibration amplitude and R&D in Applied Acoustics. He has done his doctorate in mechanical engineering from Delhi Technological University, Delhi, and M.Tech. in Machine Design from Indian Institute of Technology (IIT) Delhi. He has been a managing editor of *MAPAN-Journal of Metrology Society of India* published by Springer and a reviewer of many international journals and has enormously contributed toward evaluation, analysis, and control of noise pollution.

Dr. Mukesh Kumar is working as a dean, Science and Technology and chairperson and associate professor in the Department of Computer Science and Engineering at Gurugram University, Gurugram, Haryana. Excellence in education will be associated with Gurugram University. He received a Bachelor of Technology (B.Tech.) degree from Kurukshetra University, Kurukshetra, Haryana (India) in 2001. He then completed a Master of Technology (M.Tech.) degree from Guru Jambheshwar

University of Science and Technology, Hisar, Haryana (India), in 2004. He received a Doctorate of Philosophy in Computer Science and Engineering (Ph.D.-Computer Science Engineering) degree from Maharshi Dayanand University, Rohtak, Haryana (India) in 2012. He has published more than 150 Research papers in various national and international conferences and various peer-reviewed and referred journals of national and international reputes.

Dr. Shankar G. Aggarwal did his Ph.D. from Pt. Ravishankar Shukla University, India, in 1999. After his Ph.D. award, he worked at National Chiao Tung University, Taiwan, and at Industrial Technology Research Institute (ITRI), one of the designated institutes of Taiwan for 4 years on gas and aerosol-related research. Thereafter, he got awarded JSPS fellowship and worked at Hokkaido University, Japan, before he joined at CSIR-NPL in 2009. His current designation is a chief scientist and head of Gas Metrology of CSIR-NPL, and his current research focus is "metrology for national ambient air quality standards" in which he has been involved in gas standard preparation, air monitoring instruments related technology developments and their calibration work, etc. So far, he has been published >130 journal research papers and supervised 10 Ph.D. students. He is the member of several national technical committees of BIS, CPCB, SPCBs, NITs, NABL, etc.

Dr. Shiv Kumar Jaiswal obtained his B.E. (Electrical Engineering), M.Tech. (Electrical Engineering), and Ph.D. (Fluid Flow) degrees in 1994, 1996, and 2016, respectively. He is currently working as a chief scientist and Head—Fluid Flow Metrology Section of CSIR—National Physical Laboratory (NPLI), New Delhi. He has more than 27 years of research, industrial, and academic experience. He has been an active researcher in the field of electrical and fluid flow metrology. He has worked as a guest researcher for one year in Quantum Electrical Metrology Division of the National Institute of Standards and Technology (NIST), Gaithersburg, USA, during years 2006–07. After coming back, he established automatic 10V Josephson Voltage Standard and worked in nanometrology project for development of calibration facility for DC voltage, current, and charge measurement.

Dr. Manoj Kumar is currently a principal scientist in the Environmental Sciences and Biomedical Metrology Division at CSIR-National Physical Laboratory, New Delhi. His work spans various critical areas, including continuous emission monitoring, ambient air quality monitoring, carbon sequestration modeling, and the interaction between air pollutants and plants and microplastics. Dr. Kumar earned his M.Sc. and Ph.D. from Banaras Hindu University, Varanasi, India. Thereafter, he joined Jawaharlal Nehru University, New Delhi, as a Dr. D. S. Kothari Postdoctoral fellow. Subsequently, he joined CSIR-Institute of Himalayan Bioresource Technology in Palampur as a scientist in 2013, further enriching his experience and knowledge base. Dr. Kumar's expertise is evident in his numerous publications, which include research papers in esteemed international and national journals and contributions to book chapters. In recognition of his promising scientific endeavors, he was awarded the Start-Up Research Grant for "Young Scientists" by SERB (DST), New Delhi, in 2015.

Progress Towards the Measurement of Low Differential Pressure by Using Twin-Piston Pressure Balance in Hydraulic Mode

Chanchal, Afaqul Zafer, Renu Singh, Ajay Kumar, Raman Kumar Sharma, Lalit Kumar, and Sanjay Yadav

Abstract A system for precisely detecting the head differential pressure between a reference level and a measurement level at low line pressures has been devised. A new approach for characterizing test dial gauges is proposed, which employs a differential pressure standard in hydraulic mode within the pressure range of 0.025–0.25 MPa, with a maximum static pressure of 10 MPa. The dial gauge is characterized using a twin-pressure balance available at NPL in Delhi, India. Calibration of a dial gauge yields an approximate repeatability error of ~0.05% of readings, a maximum percentage error of 0.1–0.2% of full-scale and a calibration factor of approximately 1. The uncertainty in pressure will be calculated using a twin pressure balance with a coverage factor of $k = 1$. These characteristics will aid in the calibration performance of the twin piston pressure balance as well as in characterization of the dial gauge.

1 Introduction

Piston gauges are widely recognized as standard pressure metrology equipment. However, they are difficult to use for small differential pressure measurements since the pressure must be balanced against at least the weight of the piston, which usually

Chanchal (✉)
Government College for Women, Faridabad, India
e-mail: schanchal038@gmail.com

A. Zafer · R. K. Sharma · L. Kumar · S. Yadav
CSIR-National Physical Laboratory, New Delhi, India

A. Zafer · S. Yadav
Academy of Scientific and Innovative Research (AcSIR), Ghaziabad, India

R. Singh
BBS College of Engineering and Technology, Prayagraj, India

A. Kumar
Maharaja Surajmal Institute of Technology, New Delhi, India

S. Yadav et al. (eds.), *Advances in Metrology*, Lecture Notes in Mechanical Engineering,
https://doi.org/10.1007/978-981-96-6418-4_1

of order of several kilopascals. It is also difficult to keep the pressure fluctuations generated by piston or cylinder motion to less than 0.1 Pa [1]. To overcome these issues and measure small differential gas pressure ranges with sensitivity, a new type of piston gauge is designed. This piston gauge has two unusual features: one compensates for the piston's weight, and the other centralizes the piston in the cylinder without rotating. These systems allow the force produced due to small differential pressure produces on the piston and is measured by an electronic balance. Each pressure balance has a loaded free spinning piston in a matching cylinder. During operation, the piston floats in the cylinder due to pressure acting on its base and is balanced by gravitational force and load. The twin pressure balance generates differential pressures by using two identical balances simultaneously.

One pressure balance creates reference pressure, while the other generates both reference and differential pressure. The difference in pressure of two pressure balance gives rise to differential pressure [2, 3]. In present study, the hydraulic (sebacate oil powered) twin pressure balance utilized to calibrate the differential pressure dial gauge ranges from 0.02 to 0.25 MPa, with a maximum line pressure of 10 MPa. The dial gauge is the instrument under evaluation, and numerous parameters such as uncertainty, percentage error in pressure readings at different static pressures, repeatability, reproducibility and calibration factors are analysed to determine the dial gauge's performance [4]. The GUM (Guide to Expression of Uncertainty in Measurement) method is used as a reference for calculating uncertainty [5].

2 Experimental Technique and Calibration Procedure

Figure 1 depicts a schematic representation of a twin piston pressure balance differential pressure standard. The system consists of two mounting posts, one serving as a reference and the other as a rare one [6–8]. These mounting posts have carriage dimensions of two sets of piston cylinder assemblies, which are made of tungsten carbide and are capable of generating and measuring pressure in both hydraulic and pneumatic mode.

The static pressure range is (0.2–10) MPa and differential pressure in range (0.02–0.25) MPa is developed using stainless steel mass sets of 5 kg each. The differential pressure is always generated and measured at reference post.

The system is used to calibrate dial gauge capable to measure differential pressure from 0.2 to 0.25 MPa on static pressure from 0 to 10 MPa in steps increase of 2 MPa. After balancing the twin piston one post is isolated from the reference side and observation of dial gauge is recorded considering it as a zero differential pressure (Fig. 2).

Then the differential pressure is created at reference port in 10 steps of 0.25 MPa each until the maximum pressure of 0.25 MPa is reached. The observations is repeated 6 times at each differential pressure to make in total 60 observations at particular static pressure. Similar experiment is repeated at other steps of static pressure of (1, 3, 5, 7, 10) MPa. Hence data is recorded. Various parameters are also evaluated such as

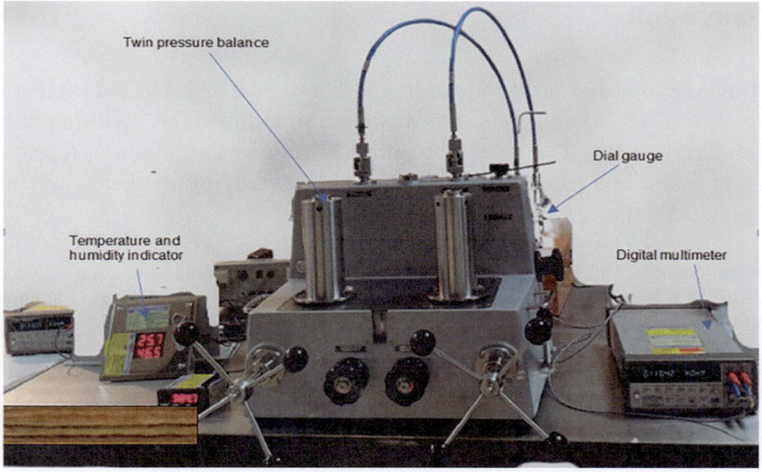

Fig. 1 Experimental setup for calibrating the dial gauge

Fig. 2 Picture of the dial gauge

Standard Deviation (MPa), Repeatability (MPa), Reproducibility, Calibration factor which will help in determining the uncertainty value [9–12].

3 Conclusion

To summarize, the dial gauge is calibrated between 0.025 and 0.25 MPa with a line pressure of 10 MPa. Calibration was done to evaluate the performance of the twin pressure balance. The percentage error in differential pressure, repeatability, calibration factor and reproducibility will help in studying the behaviour of twin piston pressure balance.

Acknowledgements The above work was done at CSIR-NPL in India. The authors are grateful to the CSIR-NPL team members for their assistance with the experimental effort. The Director of CSIR-NPL is also acknowledged for the permission to carry out this work.

References

1. Dilawar N, Varandani D, Bandyopadhyay AK, Gupta AC (2003) Characterization of a pneumatic differential pressure transfer standard. Metrologia 40:74. https://doi.org/10.1088/0026-1394/40/2/312
2. Chanchal ZA, Singh R et al (2022) A method for characterization and performance evaluation of differential pressure transducer by using twin-piston pressure balance. MAPAN J Metrol Soc India 37:379–386. https://doi.org/10.1007/s12647-021-00529-y
3. Muramoto T, Kajikawa H, Iizumi H et al. Design of a high-pressure viscosity-measurement system using two pressure balances. iopscience.iop.org
4. Singh R, Garg D (2021) Twin-piston pressure balance for measurement and uncertainty evaluation of differential pressure digital transducer. Reliab Theory Appl 16:219–225
5. Willink R (2013) Guide to the expression of uncertainty in measurement. Meas Uncertain Probab 237–244. https://doi.org/10.1017/cbo9781139135085.018
6. Rosendahl M, Nazareth RS, Magalhaes MR et al (2018) New calibration procedure for differential pressure using twin pressure balances for flowrate measurement. J Phys Conf Ser 1044. https://doi.org/10.1088/1742-6596/1044/1/012053
7. Zafer A, Yadav S (2018) Design and development of strain gauge pressure transducer working in high pressure range of 500 MPa using autofrettage and finite element method. Int J Precis Eng Manuf 19:793–800. https://doi.org/10.1007/s12541-018-0095-y
8. Yadav S, Kumar A, Dilawar N et al (2018) Role of national pressure and vacuum metrology in Indian industrial growth and their global metrological equivalence article in MAPAN. J Metrol Soc India (Springer) 33:347–359. https://doi.org/10.1007/s12647-018-0270-8
9. Bandyopadhyay AK, Olson DA. Characterization of a compact 200 MPa controlled clearance piston gauge as a primary pressure standard using the Heydemann and Welch method. 573. https://doi.org/10.1088/0026-1394/43/6/013
10. Kumar A, Yadav S, Agarwal R (2017) Design and development of a pressure transducer for high hydrostatic pressure measurements up to 200 MPa. J Inst Eng Ser C 98:413–420. https://doi.org/10.1007/s40032-016-0257-6
11. Yadav S, Prakash O, Gupta VK, Bandyopadhyay AK (2007) The effect of pressure-transmitting fluids in the characterization of a controlled clearance piston gauge up to 1 GPa. Metrologia 44:222–233. https://doi.org/10.1088/0026-1394/44/3/009
12. Chauhan J, Vijayalakshmi V, Muralidharan V, Sreedhar S (2020) Automation of hydraulic dead weight tester. In: International conference on electrical and electronics engineering, ICE3 2020, pp 236–239. https://doi.org/10.1109/ICE348803.2020.9122991

A Review on Interim Check Process of an Articulated Arm Coordinate Measuring Machine

Santosh S. Dabhole and Vinay A. Kulkarni

Abstract For real-time performance verification confidence interim checks are carried in between acceptance test and reverification test of an Articulated Arm Coordinate Measuring Machine (AACMM). Single Point Articulation Test (SPAT) is used for interim testing of AACMM which is appreciated as quick and very effective. Standards like ISO 10360-12, ASME B89.4, etc., directed interim check procedures. A review of interim check process particularly following ISO 10360-12 for AACMM with tactile sphere probing system is carried out in this paper. The various factors like sources for errors, uncertainty types, and others those are contributing to SPAT errors are studied. The product (artifact) and measurement process-related findings are presented for the scope of enhancement in quality of interim test results in view of mapping SPAT with Maximum Permissible Error (MPE) which is stated by the manufacturer.

Keywords Articulated arm coordinate measuring machine · Interim check · Single point articulation test · Tactile probing · Artifact · Error · Uncertainty

1 Introduction

1.1 An Articulated Arm Measuring Machine

An Articulated Arm Coordinate Measuring Machine (AACMM) known for its excellent utility value in Micro, Small and Medium Enterprises (MSMEs) is becoming popular because of its capability of providing measurement results in prompt manner which saves large amount of job handling time and its associated inventory cost. In real time on the shop floor, the measurement tasks which are carried out by using

S. S. Dabhole (✉) · V. A. Kulkarni
DY Patil College of Engineering Akurdi, Savitribai Phule Pune University, Pune, India
e-mail: santosh1152@rediffmail.com

S. S. Dabhole
Nutan Maharashtra Institute of Engineering and Technology, Talegaon Dabhade, India

S. Yadav et al. (eds.), *Advances in Metrology*, Lecture Notes in Mechanical Engineering,
https://doi.org/10.1007/978-981-96-6418-4_2

AACMM are not always in ideal environment that are there inside the laboratory conditions.

In Fig. 1, six rotary axes AACMM is shown at top and seven rotary axes at bottom. It has mechatronics system elements [2] and cyclic variation in the temperature in test area, magnitude of vibration present, humidity variations, cleanliness, operator-to-operator variation in operative skills of measuring equipment handling, operator's perception, and level of expertise are other variables in its use. There are other contributors also, these contributors in combination adds doubt on extent on which these are considerably taken as reliable. When AACMM are used in non-standard conditions care must be taken to achieve reliable and there must be less amount of uncertainty associated in the measurands. Any abrupt handling of AACMM's parts like probing system, stylus, probing over pressure, arm, etc., must be avoided for quality of measurement results.

As a standard practice Coordinate Measuring Machines (CMMs), Micro-CMM must have their calibration process or reverification done for ensuring the higher degree of confidence in the measurement during their effective service span [3]. There are some cases wherein along with physical standards the light source like laser is used as length standard [4].

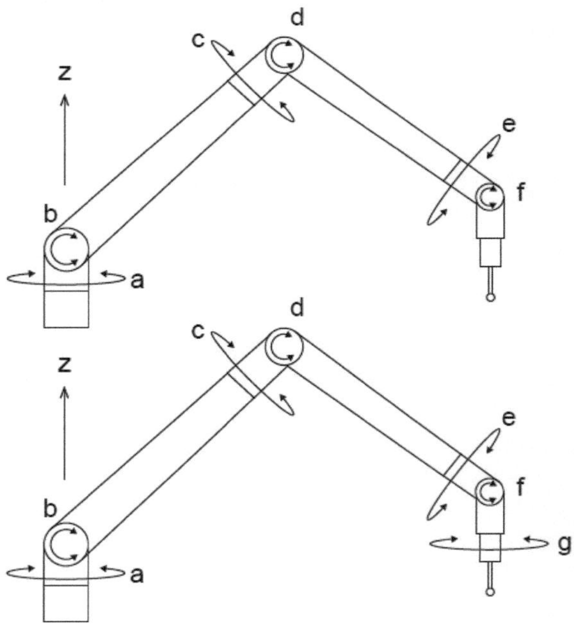

Fig. 1 Schematic of AACMM [1]

1.2 Significance of Study

For AACMM, ISO 10360-12 is known as standard for Geometrical Product Specifications (GPS)—Acceptance and reverification tests for coordinate measuring systems (CMS)—Part 12: Articulated arm coordinate measurement machines (CMM). Its Annex D the interim test recommends Single Point Articulation Test (SPAT) as an effective test that helps user to rate real-time performance of AACMM for its given results in terms of quality and reliability. Interim test described uses kinematic nest artifact for verification of SPAT error. In the ideal form kinematic nest is a rigid assembly which uses each of the three spherical ball's point of contact with measuring probe sphere simultaneously to form a seat for measuring probe position.

In the interim test procedure explained in the standard for SPAT error evaluation, there are not any specific guidelines mentioned for selection of the probe sphere space point location or region in the range envelope of an AACMM, however ASME B89.4 suggested these sampling to be done at three spot locations which are in the specified range of AACMM for interim test.

The choice in selection of specific region or location within the range envelope of AACMM may have significant impact on interim test results as AACMM is built with complex electromechanical structural joints and in its construction, there is an incorporation of various electrical, sensory, mechanical, mechatronics systems. These elements used come with their inherent characteristics that leads to have an impact on digitally produced measurement results. The quality personnels have their own choice to decide the location of the SPAT error verifications point. In this regard, to assure the reliability in measurement results and the enhanced quality confidence level of AACMM results there is a scope for research for finding the probing point region or locations in a three-dimensional space envelope of AACMM in interim test. It must be clear that interim test is neither a calibration test nor a verification test. It finds its scope in the quality assurance section where the degree of quality of the measuring system in particular AACMM is set between the due period of calibration as per previous calibration traceability report of measuring equipment. This test finds its usefulness to stop using AACMM which has degraded performance at the stage to avoid rejection of good parts if faulty measuring machine is being used. In this study SPAT error contributing factors are studied.

2 Literature Survey

AACMM has its excellent portability features, greater volumetric compass and is a speedy precision measurement machine [5]. There exist a few limitations also due to random errors [6]. For heavy job's real-time feature inspection tasks at shop floor AACMMs with touch probing systems are becoming popular as it offers great portability feature. Outmost care must be taken while handling of the AACMM's parts during the measurement process as the factors like changes in external environmental conditions like temperature, extent of humidity, vibrations of parts, and dusts cause variation in results. The intrinsic factor like positioning of the rotary encoders at joints as a mechatronics element also contributes to error component in measurands [2, 7]. These electronics and electrical elements are key constituents for AACMM as they are working in sensory functions. As an electronic element encoder has inherent uncertainty characteristics which has always an error in the AACMM results [6, 8]. In an Industry 4.0 perspective in each phase of the product lifecycle span traceable measurement has its own impact in current swift design to manufacturing processes [9]. The AACMM gives more part of error component in comparison with bridge type CMMs. In between the two periods of calibration of AACMM metrologist using the AACMM needs to be very attentive and propose a good degree of belief that results of measurement achieved are at highest degree of confidence [10]. The international standards more popular and used for performance verifications of portable CMM are, viz., VDI 2617.9-2009, ASME B89.4.22, ISO 10360-12 [1, 3, 11–14].

As an industrial practice to verify coordinate measuring machine's drift response, a calibrated artifact and statistical methods are used. The high-speed sintering additive manufactured artifact developed with various geometrical features are becoming popular. These features are measured using tactile methods, non-tactile methods like optical, and X-ray computed tomography by coordinate measuring systems are also becoming popular. The results of these methods highlighted difference between these measuring methods. There exist need to focus on development for specification standards and guidance for the dimensional assessment of additive manufacturing parts. It signifies the need for specification standards and set guidelines for dimensional assessment for benchmarking artifacts. A benchmarking artifact is made and following the "design-for-metrology" approach quality control optimization is done [15].

Recently additive manufactured artifacts are becoming popular. In the design stage, the uncertainty evaluation must be considered when designers design the additive manufactured artifact. An additive manufactured artifact developed has a limitation if boss-like feature is present in it and it needs to analyzed carefully for redesign of it [15, 16]. Attention was drawn to the constraint of establishing standardization in interim check devices by Arenhart et al. [10]. Their review findings recommended additional research, development, and verification techniques for interim check systems. For precise and tight measurement results touch trigger probe is found to be superior to scanning method of probing. Every method of inspection by probing has its advantages and constraints [17]. The portable CMM normally uses touch trigger method of probing. Some machines have capacity for scanning probing which is useful for non-contact measurement tasks [18].

For different kinds of metrology domains, there is a design and development challenges exist for development of low-cost accurate sensory devices which consume less power [8]. The artifact measurement results are always associated with the errors with the uncertainty [19]. Calibration certificate clarifies the uncertainty part that must be taken in account of measurement result after the measurement is carried out by respective measurement instrument [20]. To design an interim check device, standards such as ISO 10360-2 and NPL's Good Practice Guide No. 42 are helpful

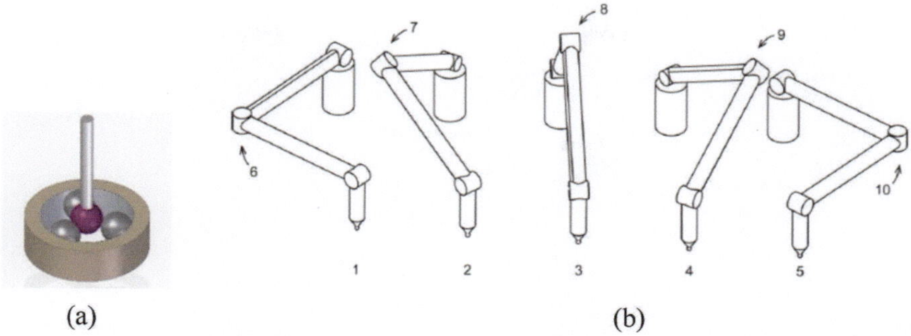

Fig. 2 SPAT measurement: **a** regular three ball trihedral set, **b** elbow positions

[1, 10, 21]. Trihedral seat also known as kinematic nest comprising three hard balls with ring (see Fig. 2) is used in performance checking of AACMM [1, 14]. The focus of research work is the trihedral seat and its related measurement process for interim check, see Fig. 2.

The development in this kinematic seat system is required [22]. The electronics position sensors used in this kinematic nest add extra cost to measurement setup for SPAT test. The uncertainty work is not evaluated out for this digital nest [7]. For PCMM measurements, it is frequently seen that measurement value at the end of measurement process varies from operator to operator. The literature studied in this regard considers standard uncertainty that is always associated with the measurands [2, 4, 5, 9, 13, 14, 18, 20]. For the task-specific measurements carried out by using AACMM, the expanded uncertainty estimations are also calculated for some of the works. The fishbone diagram (see Fig. 3) shows the various elements associated with Single Point Articulation Test (SPAT) error observed in AACMM. The variables like measuring machine, workpiece, operator, measurement strategy, and environmental factors are major contributors for SPAT error.

2.1 Field Survey

In interaction with few of AACMM inspection service providers in Maharashtra Industrial Development Corporation (MIDC) Bhosari, Pune, it is observed that the calibration expenses for a single AACMM costs an approximately Fifty Thousand Indian Rupees, which may vary from place to place. The scope for developing low-cost interim check device is possible for AACMM which in turn useful for Quality department to increase the confidence level of measurement process.

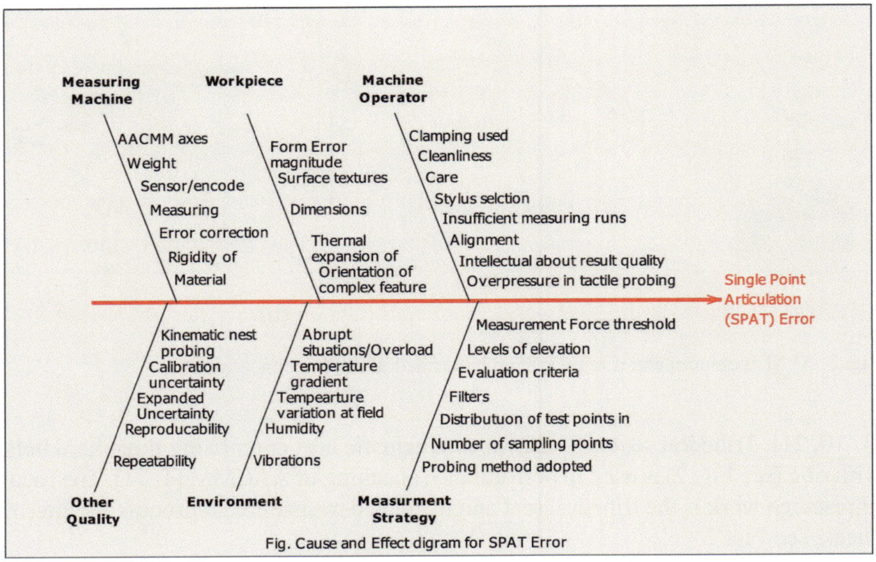

Fig. 3 SPAT error fishbone diagram

2.2 *Present State of Art*

AACMM is also known as Portable Coordinate Measuring Machine (PCMM). This machine in construction have various joints for articulation of the arm and other elements of PCMM. This makes constraints and low repeatability results achieved by PCMM than the precise bridge type CMM. PCMM has the advantage for measurement tasks for the complex and larger size jobs at shop floors. It is known that the tests like standard calibration, acceptance test, and reverification tests are done at specified laboratory conditions. The interim tests are optional for the organization who use AACMMs. ISO 10360-12, ASME B89.4 give directions and procedures about how the plan of measurement process for the interim tests. These standards demonstrate standard practices and use of various artifacts used for interim test. The Single Point Articulation Test (SPAT) in ASME B89.4.22 is used for spot verification testing of AACMM for its reliability. Recent days ISO 10360-12 is followed by various Portable CMM manufacturing industries. In India IS 15635 (Part 12): 2023 is used which is identical to ISO 10360-12:2016. Annex D of this standard which is informative, gives interim test parameters and procedures. This study is made for comprehensive review of this standard along with available literature for interim test artifact design and development for AACMM.

2.3 Single Point Articulation Test (SPAT)

Principle: An articulated arm's correct operations and compensations are quickly and effectively verified by SPAT. The variation in the spherical tactile probe positions is evaluated by use of a kinematic seat. There are artifact types to choose from few of them such as three ball kinematic nest, chamfered hole seat, conical socket with cone angle from 60° to 120°. Kinematic mounting seat also known as a trihedral socket seat is most effective among these artifacts. The elbow of the AACMM is articulated about the kinematic nest seat in a defined way as per written procedure in standard. Figure 2 shows the regular three hardened ball kinematic nest used for SPAT test with ring. This set up is used for repeatability test of measuring sphere probe. This measurement process has ten readings and average is calculated. The maximum gap difference between measured readings and the average of ten is considered as SPAT error. It is important that the absolute value of SPAT error must be less than the manufacturer provided value.

Measuring Equipment

For this study only kinematic nest is considered as an interim test artifact. Kinematic seat formed may be formed with a conical socket or trihedral ball socket with a magnet attached in bottom of set to ensure that a gentle tactile probing on seat is created. For SPAT the qualification test criteria of the measuring equipment is the repeatability of positioning sphere is within the 20% value of the maximum that is recommended. The standard also recommends larger diameter sphere is placed over the trihedral nest. This standard also recommended the conical socket having cone angle between 60° and 120°. It also specifically instructed that one-third of probe diameter should be inside the cone.

Procedure

Section D.2.3 describes the total ten points readings are accounted for stated elbow articulations by ensuring the probe sphere makes gentle contact with kinematic seat formed with trihedral nest (see Fig. 2b). As per ISO 10360-12 standard measurement at any point location within the volumetric range of AACMM can be selected for SPAT verification.

Derivation of Test Results

For the probe tip Average Cartesian coordinate is calculated by average of these ten measured coordinate values. Along X, Y, and Z directions half the range is calculated for each respective directions. The maximum of these three calculated range values is the SPAT error. SPAT error thus calculated if exceeds than maximum permissible error (MPE) value which is given by the manufacturer, then focus must be made to do corrective measures. It is very important to investigate and do root cause analysis in AACMM or in the measurement process, setup conditions, etc.

Error and Uncertainty in CMM Results

In dealing with measurement results important aspects needs to be considered like errors and uncertainties. In metrology error is the gap between the value of measurand achieved at measurement and the true value of that feature. For every measurand uncertainty is also a part of measured value.

Probing Errors

Excessive probing forces make deformations in mechanical elements of AACMM's probing system. It also causes certain deflection in the ball bar, fixture elements of measurement setup. Ball bar positions susceptible to vibrations must be avoided. The reason for vibration generation can be external or internal. AACMM operator needs to take care that sufficient vibration dampening measures in the measurement set up are incorporated to reduce errors that are caused because of it. The excessive overhangs of arm member for longer time must be avoided during inspection. Sagging shape deformation caused due to gravity effect in vertical position shortens the length when it is held horizontally in performance test of AACMM. Also there is a cosine error formed. Overpressure must not be applied by the operator over the artifact surfaces. Hertzian stresses are developed due to interaction of surfaces with increased contact forces. At certain levels these plastic deformations are formed on the probe surfaces or artifact surfaces, and this has critical impact in generating errors in the measured values for repeated readings.

Uncertainty

Uncertainty is a parameter, associated with the result of a measurement, i.e., measurand value that characterizes the dispersion of the values that could reasonably be attributed to the measurand [23, 24].

The mathematical model for measurand is expressed as

$$Y = x_i + \Delta x \pm U \tag{1}$$

where Y = true value of measurement quantity, x_i = actual inspected dimension, Δx = errors, and U = expanded uncertainty.

Expanded uncertainty is used to express the percentage of the confidence interval for the measurement result within which the true value of the measurand is believed to lie. Expanded uncertainty is calculated by multiplying the combined uncertainty with coverage factor (k) [24].

The combined uncertainty is derived from Type-A and Type-B uncertainties. The coverage factor (k) is calculated from the student's t distribution for respective effective degrees of freedom and confidence level.

The combined uncertainty σ_Y is obtained by the root mean square method given by

$$\sigma_Y = \sqrt[2]{f_{x1}^2\sigma_{x1}^2 + f_{x2}^2\sigma_{x2}^2 + \ldots f_{xn}^2\sigma_{xn}^2} \tag{2}$$

where f_{x1}, f_{x2} …, f_{xn} representing partial derivatives of respective measurement model. This model represents the relativity between the input quantities and the quantity being measured. The partial derivatives are also used as sensitivity coefficients. These sensitivity coefficients give the effects of each input variable on the final results. In other terms, it describes the sensitivity of the output quantity with respect to each of the input quantity. In the above equation, $\sigma_{x1}, \sigma_{x2}, \ldots, \sigma_{xn}$ represents the standard uncertainties that are obtained from Type-A and Type-B method.

$$U = k\sigma_Y \quad (3)$$

where k is a coverage factor, it is a numerical factor used to calculate an expanded uncertainty as a multiplier of the combined standard uncertainty. σ_Y is the combined uncertainty [24].

Type-A Uncertainty Evaluation

For the same metrological conditions standard uncertainty is calculated for repetitive measurements. Gaussian or normal distribution is followed by the frequency plots of the measurements. If x_1, x_2, …, x_n are the measurement values recorded for n repeated measurements then standard uncertainty σ_{x1} is given by

$$\text{Standard uncertainty } (\sigma_{x1}) = \frac{\sqrt{\text{Var}(x)}}{\sqrt{n}} \quad (4)$$

where , $\overline{x}$ = mean, and n = number of measurements.

Type-B Uncertainty Evaluation

Type-B uncertainty is that which is evaluated other than statistical method of evaluation which includes the data obtained from the manufacturer's specifications, scientific judgment, domain experience conducted for series of readings [23]. There is a probability distribution function (PDF) which is characterized with the Type-B uncertainty contributing elements. Therefore, all these mathematical functions need to be normalized by using some correction factor to a standard uncertainty as following.

1. Normal distribution (2 for 95% confidence level, 3 for 99% confidence level)
2. U shaped distribution ($\sqrt{2}$)
3. Triangular distribution ($\sqrt{6}$)
4. Rectangular distribution (($\sqrt{3}$)

3 Limitations of State of Art

3.1 *Artifact Related Limitations*

The artifacts used for interim tests for AACMM generally formed with kinematic seat developed from trihedral set or conical sockets. This consists of three hard balls placed inside ring. These three balls make stable contact with the sphere of AACMM which comes down from top. At an measurement instant these three balls need to come in gentle contact with measuring sphere. The probing pressure affect the accuracy of correct measurement results. The excessive probing pressure develops friction between the materials which are similar or dissimilar. This creates force on inside cylindrical walls of ring which hoses three balls. The difference in surface hardness between the measuring elements like sphere surface and probe tend to have an error contribution in measurement results. As operator to operator handling of AACMM is done at variable skill level, above points are associated with material wear. In perspective view it has tribological importance too. Care must be taken for avoiding probing over force and in real time gentle touch of spheres must be ensured to better coordinate reading by AACMM. Equiangular positions of three balls in kinematic nest also ensure this to have a quality of recorded points. There is need to work on above issues for always having a quality of measurement results in interim test. As interim test is a self-check method of verification and not a substitute to calibration. A reliable interim test artifact and method needs to be developed.

3.2 *Interim Check Process Related Limitations*

The measurement process for interim check recommended by ISO 10360-12 in its Annex D states that any point location within the measuring volume of AACMM may be selected for measurement of SPAT test. ASME B89.4.22 clearly defined three sampling points location in the measuring range of AACMM as qualification for interim test result [25]. The specific measuring range of 60–100% for AACMM is used for testing of calibration length. Calibrated length artifact ensures covering minimum 66% of measuring range length of AACMM for verification. This indicates that as you articulate the measuring arms from the base of the AACMM toward the extreme range of the AACMM more impact on reliability of measurement results observed. AACMM in its extreme range articulation experiences complex configuration of loads on joints. AACMM uses kinematic transformations for its results. It also has complex mechatronics sensory system also. Therefore, articulation of the arms in extreme range region generates more magnitude of errors as compared to the position of arm which is nearer to the center of gravity of AACMM. This has impact on error value in the measurement results. The point selection criteria for interim test is not described in ISO 10360-12. In the ASME B89.4.22, it has clear description of the three points sampling locations selection in AACCM's range. For performing interim test at highest quality level of confidence of the metrologist needs to be develop qualification criteria of selected point/s for SPAT process.

4 Conclusion

In this paper, a review of interim test as per ISO 10360-12 or other standards for test artifact development and related process is done which uses contact probing. The scope of interim test is to only ensure fitness qualification of the AACMM in real-time basis and not as a substitute of calibration process is realized. The quality of measurements results conducted for SPAT error verification has several connections like operator, environment, strategy, machine, an artifact, strategy, uncertainties, etc. Maximum Permissible Error value which is provided by a AACMM manufacturer needs to be referred along with SPAT error. This study finds need of improvement in strategy for quality in selection of test point location and also attract need of digital artifact for the purpose of interim test. The artifact and the optimum process for SPAT error verification can be worked out for reliable results with superior quality of inspection data and saves cost through the interim check process.

References

1. ISO 10360-12 for Geometrical Product Specifications (GPS). Acceptance and reverification tests for coordinate measuring systems (CMS)—Part 12: Articulated arm coordinate measurement machines (CMM)
2. Cuesta E, Alvarez B, Sanchez-Lasheras F, Gonzalez-Madruga D (2015) A statistical approach to prediction of the CMM drift behaviour using a calibrated mechanical artefact. Metrol Meas Syst (Walter de Gruyter GmbH) 22(3):417–428. https://doi.org/10.1515/mms-2015-0033
3. Claverley JD, Leach RK (2015) A review of the existing performance verification infrastructure for micro-CMMs. Precis Eng (Elsevier BV) 39:1–15. https://doi.org/10.1016/j.precisioneng.2014.06.006
4. Ostrowska K, Gąska A, Kupiec R, Sładek J, Gromczak K (2016) Verification of articulated arm coordinate measuring machines accuracy using laser tracer system as standard of length. MAPAN (Springer Science and Business Media LLC) 31(4):241–256. https://doi.org/10.1007/s12647-016-0176-2
5. Gąska P, Gąska A, Sładek J, Jędrzejewski J (2019) Simulation model for uncertainty estimation of measurements performed on five-axis measuring systems. Int J Adv Manuf Technol (Springer Science and Business Media LLC) 104(9–12):4685–4696. https://doi.org/10.1007/s00170-019-04319-4
6. Rim C-H, Sun B-Q, Kim Y-G, Kim P (2019) Analysis of random factors affecting measurement accuracy of portable coordinate measuring arm. MAPAN (Springer Science and Business Media LLC.) 34(4):529–539. https://doi.org/10.1007/s12647-019-00323-x
7. Ibaraki S, Saito R (2023) Novel kinematic model of articulated arm coordinate measuring machine with angular position measurement errors of rotary axes. CIRP Ann (Elsevier BV). https://doi.org/10.1016/j.cirp.2023.03.035

8. Chaudhary KP, Dubey PK, Gahlot A, Dahiya A (2021) Advances in sensors and measurements for metrological applications. MAPAN (Springer Science and Business Media LLC) 36(2):211–213. https://doi.org/10.1007/s12647-021-00482-w
9. Carmignato S, De Chiffre L, Bosse H, Leach RK, Balsamo A, Estler WT (2020) Dimensional artefacts to achieve metrological traceability in advanced manufacturing. CIRP Ann (Elsevier BV) 69(2):693–716. https://doi.org/10.1016/j.cirp.2020.05.009
10. Arenhart RS, Pizzolato M, Menin PL, Hoch L (2021) Devices for interim check of coordinate measuring machines: a systematic review. MAPAN (Springer Science and Business Media LLC) 36(1):157–173. https://doi.org/10.1007/s12647-020-00406-0
11. Acero R, Brau A, Santolaria J, Pueo M (2017) Evaluation of a metrology platform for an articulated arm coordinate measuring machine verification under the ASME B89.4.22-2004 and VDI 2617_9-2009 standards. J Manuf Syst (Elsevier BV) 42:57–68. https://doi.org/10.1016/j.jmsy.2016.11.002
12. Zhao H, Yu L, Xia H, Li W, Jiang Y, Jia H (2018) 3D artifact for calibrating kinematic parameters of articulated arm coordinate measuring machines. Meas Sci Technol (IOP Publishing) 29(6):065009. https://doi.org/10.1088/1361-6501/aab807
13. Rim C-H, Rim C-M, Kim J-G, Chen G, Pak J-S (2018) A calibration method of portable coordinate measuring arms by using artifacts. MAPAN (Springer Science and Business Media LLC) 34(1):1–11. https://doi.org/10.1007/s12647-018-0297-x
14. Mutilba U, Kortaberria G, Olarra A, Gutiérrez A, Gomez-Acedo E, Zubieta M (2013) Performance calibration of articulated arm coordinate measuring machine. Procedia Eng (Elsevier BV) 63:720–727. https://doi.org/10.1016/j.proeng.2013.08.264
15. Rivas Santos VM, Thompson A, Sims-Waterhouse D, Maskery I, Woolliams P, Leach R (2020) Design and characterisation of an additive manufacturing benchmarking artefact following a design-for-metrology approach. Addit Manuf (Elsevier BV) 32:100964. https://doi.org/10.1016/j.addma.2019.100964
16. Tanveer MQ, Suhaib M, Haleem A (2020) A new 3D benchmarking artifact to evaluate dimensional accuracy and geometric tolerancing of additive manufacturing technique. In: Lecture notes in mechanical engineering. Springer Singapore, pp 261–273. https://doi.org/10.1007/978-981-15-1071-7_22
17. Rupal BS, Ahmad R, Qureshi AJ (2018) Feature-based methodology for design of geometric benchmark test artifacts for additive manufacturing processes. Procedia CIRP (Elsevier BV) 70:84–89. https://doi.org/10.1016/j.procir.2018.02.012
18. Bastas A (2020) Comparing the probing systems of coordinate measurement machine: scanning probe versus touch-trigger probe. Measurement (Elsevier BV) 156:107604. https://doi.org/10.1016/j.measurement.2020.107604
19. Sepahi-Boroujeni S, Mayer JRR, Khameneifar F (2021) Efficient uncertainty estimation of indirectly measured geometric errors of five-axis machine tools via Monte-Carlo validated GUM framework. Precis Eng (Elsevier BV) 67:160–171. https://doi.org/10.1016/j.precisioneng.2020.09.027
20. Brosed FJ, Acero R, Aguado S, Herrer M, Aguilar JJ, Santolaria J (2019) Design and development of a calibration artefact for length measurement system. Procedia Manuf (Elsevier BV) 41:58–65. https://doi.org/10.1016/j.promfg.2019.07.029
21. Flack D. Good practice guide No. 42. CMM verification. Issue 2 Engineering Measurement Division National Physical Laboratory
22. Arenhart RS, Pizzolato M, Beuren FH, Souza AM, da Rosa LC (2022) device for interim check of coordinate measuring machines. Metrol Meas Syst 29(1):143–158. https://doi.org/10.24425/mms.2022.138546
23. Bell S. Measurement good practice guide No. 11 (Issue 2)—a beginner's guide to uncertainty of measurement. ISSN 1368-6550
24. JCGM (2008) 100:2008, GUM 1995 with minor corrections. Evaluation of measurement data—guide to the expression of uncertainty in measurement, 1st edn. Corrected version 2010
25. ASME B89.4.22-2004 methods for performance evaluation of articulated arm coordinate measuring machines

Analysis of Source and Source Region of Coarse Mode Aerosol (PM_{10}) in Varanasi, India

Preeti Tiwari, Bharat Ji Mehrotra, Manoj K. Srivastava, Manoj Kumar, Narayanasamy Vijayan, and Sudhir Kumar Sharma

Abstract In this study, we conducted source apportionment and source region analysis of elemental concentrations of coarse mode aerosol (PM_{10}) using principal component analysis (PCA), trajectory analysis, and conditional bivariate probability function (CBPF) in Varanasi, situated over central Indo-Gangetic Plain (IGP) of India. In this case, 34 elements (Mg, Na, Al, Si, K, Ca, Ti, Cr, Mn, Fe, Ni, Cu, Zn, Ga, Zr, Mo, Ag, Pm, Y, Rb, Pd, Fr, U, Pb, B, P, S, Cl, Br, Nb, Sr, F, As, and Ba) were extracted from PM_{10} samples using WD-XRF technique. PCA identified six sources (crustal/soil/road dust, biomass burning, industrial emissions, agricultural activities, combustion, and vehicular emissions) of elemental concentrations of PM_{10} over Varanasi. Trajectory analysis indicated contribution to PM_{10} concentration transported from Uttar Pradesh, Punjab, Haryana, Delhi, partially from Rajasthan, Gujarat, West Bengal, and some parts of Madhya Pradesh, with additional transboundary emissions from Pakistan, Afghanistan, and Nepal. CBPF analysis predicted the local PM_{10} emissions from North-East, South, West, and North-West directions, with major contributors including Ganga ghats, religious places, market areas, and traffic activities.

Keywords PM_{10} · Source apportionment · Source region · Trajectory analysis · CBPF

P. Tiwari (✉) · M. Kumar · N. Vijayan · S. K. Sharma
CSIR-National Physical Laboratory, New Delhi, India
e-mail: preetis2605@gmail.com

Academy of Scientific and Innovative Research (AcSIR), Ghaziabad, India

S. K. Sharma
e-mail: sudhircsir@gmail.com

B. J. Mehrotra · M. K. Srivastava
Department of Geophysics, Banaras Hindu University (BHU), Varanasi, India

S. Yadav et al. (eds.), *Advances in Metrology*, Lecture Notes in Mechanical Engineering,
https://doi.org/10.1007/978-981-96-6418-4_3

1 Introduction

The Earth's atmosphere contains a small proportion of atmospheric aerosols with brief residence times. Despite their limited presence, these aerosols play a crucial role in modifying the regional atmospheric radiation budget. They accomplish this both directly, by absorbing or reflecting solar energy, and indirectly, by influencing the formation of clouds [1, 2]. The transit of polluted air masses from densely populated areas in South Asia over considerable distances while retaining their distinct characteristics is a notable phenomenon [3]. These air masses are laden with high concentrations of carbonaceous aerosols, which are predominantly released from the combustion of bio-fuels and fossil fuels [4]. The swift pace of urban expansion, industrial growth, and the escalating volume of vehicular traffic, especially within the Indo-Gangetic Plain (IGP) region, significantly add to the mounting burden of atmospheric particulate matter [5–7]. Originating from the IGP, a key source region, these pollutants not only degrade local air quality but also undergo extensive long-distance transport, facilitated by the prevalent convective activity characteristic of tropical regions [8].

During the summer months (March-June), the extensive dispersal of mineral dust from regions including Iran, Afghanistan, Pakistan, and the Thar Desert in western India introduces notable fluctuations in aerosol composition over long distances [9, 10]. Additionally, the amalgamation of carbonaceous compounds with anthropogenic inorganic elements has resulted in deteriorated air quality, reduced visibility, and formation of atmospheric brown clouds during winter (December-February) [11, 12]. The transportation and presence of pollutants across the Himalayan region, and their influence on monsoon circulation and rainfall patterns across India, have sparked significant debate and scrutiny [13].

The current research aims to examine the fluctuations in chemical components (especially major and trace elements) of PM_{10} within the central region of the IGP, specifically focusing on Varanasi, from January 2019 to December 2019. The study entails a comprehensive analysis of elemental constituents and the attribution of PM_{10} sources using the PCA-APCS receptor model. Through trajectory analysis, the study identifies both local and transboundary movements of PM_{10} from their origins to the receptor site. This investigation promises to offer valuable insights into the sources, source region, and transport routes of PM_{10}, thereby enhancing our understanding of its dynamics.

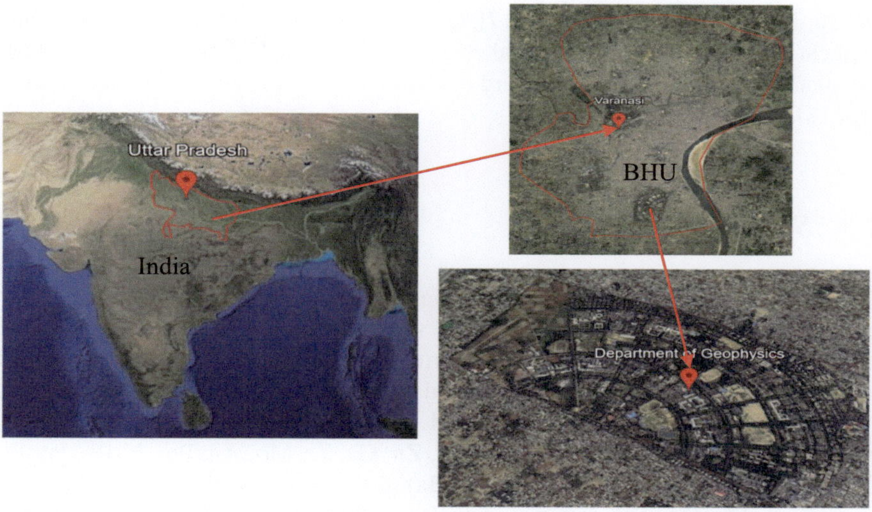

Fig. 1 Sampling site of PM_{10} at Banaras Hindu University, Varanasi. *Source* Google Earth

2 Methodology

2.1 Sampling Site

PM_{10} samples were collected at Banaras Hindu University (BHU) in Varanasi, India (25°26′N, 82°99′E; 80 m amsl) from January to December 2019 (Fig. 1). The sampling was conducted using 8″ × 10″ pre-combusted quartz microfiber filters. A high-volume respirable dust sampler (AAS 217 NL, Ecotech, India) was used for the sampling, operated at an average flow rate of 1.12 $m^3\ min^{-1}$ with an accuracy of ±2%. The sampler was positioned approximately 10 m above the ground (AGL). The initial and final weights of the PM_{10} filters were measured using a microbalance with a resolution of ±10 μg, calibrated with traceable standards. The concentration of PM_{10} was determined using the standard gravimetric method [14–16].

2.2 Chemical Analysis

Elemental analysis of PM_{10} samples was conducted using a Wavelength Dispersive X-Ray Fluorescence (WD-XRF) Spectrometer (Rigaku ZSX Primus model). The instrument featured a scintillation counter (SC) for heavy elements, a flow-proportional counter (F-PC) for light elements, and a Rh-target X-ray tube operating at 2.4 kW under vacuum at 36 °C [17]. Different analyzer crystals were used for element detection: RX25 for Mg, PET for Al, Ge for P, S, and Cl, and LiF (200) for Ca, K, Ti, Cr, Mn, Fe, and Zn [17]. QA/QC involved field blank corrections and triplicate analysis, with an analytical error of 5–10%. Method detection limits were calculated as three times the standard deviation of 10 replicate field blank samples [15, 18].

2.3 Source Apportionment

Principal Component Analysis (PCA) was utilized to identify potential sources of PM_{10} mass at the selected site. This statistical tool detects patterns in data, highlighting their differences and similarities. In this study, PCA was conducted using an orthogonal transformation method with Varimax rotation. The analysis was constrained to factors with eigenvalues >1. A total of 34 constituents/species (Mg, Na, Al, Si, K, Ca, Ti, Cr, Mn, Fe, Ni, Cu, Zn, Ga, Zr, Mo, Ag, Pm, Y, Rb, Pd, Fr, U, Pb, B, P, S, Cl, Br, Nb, Sr, F, As, and Ba) of PM_{10} were included as variables in the dataset. The chemical data were transformed into a dimensionless standardized form for the PCA.

$$Z_{ij} = \frac{C_{ij} - C_j}{\sigma_j} \tag{1}$$

where $i = 1, \ldots, n$ samples; $j = 1, \ldots, m$ elements; C_{ij} is the concentration of element j in sample i; and C_j and σ_j are the arithmetic mean concentration and the standard deviation for element j, respectively. The PCA model is expressed as

$$Z_{ij} = \sum_{k=1}^{p} g_{ik} h_{kj} \tag{2}$$

where $k = 1, \ldots, p$ sources and g_{ik} and h_{kj} are the factor loadings and the factor scores, respectively. Eigenvector decomposition helps in solving this equation [19].

Source profiles and source contributions were then quantitatively estimated based on the factor loading scores from PCA, using the Absolute Principal Component Score (APCS) method as described by Thurston and Spengler [20] and Henry and Hidy [21]. Since the data for PCA results were normalized, the true zero for each factor score was derived accordingly.

2.4 Source Region Analysis

2.4.1 Backward Trajectory Analysis of Air Masses

Backward trajectory analysis of air masses was conducted using datasets from the Air Resource Laboratory (ARL) under the National Oceanic and Atmospheric Administration (NOAA). The Hybrid Single Particle Lagrangian Integrated Trajectory (HYSPLIT) model available at (http://ready.arl.noaa.gov/HYSPLIT.php) by employing GDAS global meteorological data was utilized to trace the air mass pathways of particulate matter (PM_{10}) from the receptor site, namely, Varanasi. TrajStat software was used to generate and analyze 72 h backward trajectories at 500 m AGL for each sampling day from January 2019 to December 2019 [22].

2.4.2 Conditional Bivariate Probability Function (CBPF)

The conditional bivariate probability function (CBPF) extends the conditional probability function (CPF) by incorporating wind speed as a third variable, distributing pollutant concentrations across cells defined by both wind direction and speed. This approach, as detailed by Uria-Tellaetxe and Carslaw [23], utilizes polar coordinates to better identify source regions. In this study, the Open-air package in RStudio was used to implement CBPF, with meteorological data, including wind speed and direction, collected from automatic weather stations since 2019 [24]. The analysis focused on PM_{10}, considering the 75th percentiles using the polar plot function [22].

3 Results and Discussion

3.1 Particulate Matter Concentration

The annual average mass concentration of PM_{10} with standard error at BHU, Varanasi, was reported to be $228 \pm 34\ \mu g\ m^{-3}$, which was approximately 3.8 times higher than the annual average concentration of National Ambient Air Quality Standards (NAAQS) for PM_{10} ($60\ \mu g\ m^{-3}$ annual), and 15.2 times higher than the World Health Organization (WHO) limits ($15\ \mu g\ m^{-3}$). Figure 2 shows the temporal variation of PM_{10} concentration at the study site. Elevated levels of PM_{10} have been associated with adverse health effects, such as impaired lung function and growth in children, along with respiratory issues and asthma exacerbations. Furthermore, adults exposed to high concentrations of PM_{10} may be at increased risk for developing heart diseases, diabetes, and neurological problems [17, 25].

Previous research in Varanasi and adjacent locations within the IGP reported average PM_{10} concentrations of $99 \pm 33\ \mu g\ m^{-3}$ in Varanasi, $135 \pm 64\ \mu g\ m^{-3}$ in Delhi, and $116 \pm 38\ \mu g\ m^{-3}$ in Kolkata [14]. Additionally, another study in Delhi

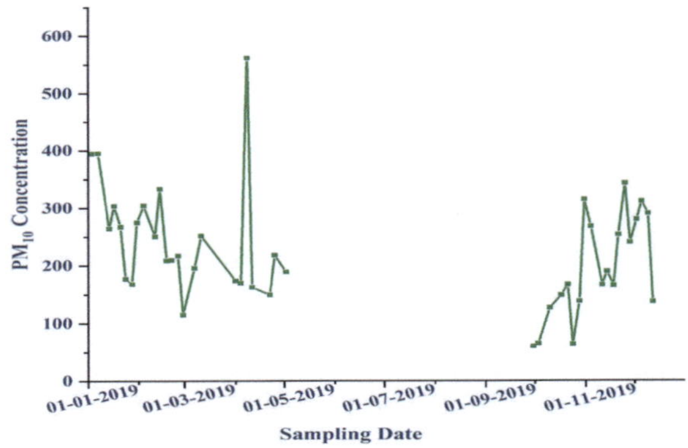

Fig. 2 Temporal variation of PM_{10} concentration at sampling site

and Faridabad found PM_{10} concentrations in the IGP region to be 209 ± 81 $\mu g\ m^{-3}$ and 195 ± 121 $\mu g\ m^{-3}$, respectively [26].

3.2 *Elemental Constituents*

The annual average elemental concentration of PM_{10} in Varanasi is shown in the figure below. Identification of all 34 elements (Mg, Na, Al, Si, K, Ca, Ti, Cr, Mn, Fe, Ni, Cu, Zn, Ga, Zr, Mo, Ag, Pm, Y, Rb, Pd, Fr, U, Pb, B, P, S, Cl, Br, Nb, Sr, F, As, and Ba) was conducted using WD-XRF technique (Fig. 3).

Among all the elements in PM_{10} mass load, the predominant elements included Si (13.7 ± 2.07), Al (5.5 ± 0.08 μgm^{-3}), S (4.84 ± 0.73), Ca (4.27 ± 0.64), K (3.82 ± 0.57), Fe (3.08 ± 0.46), Cl (1.98 ± 0.3), and Mg (1.13 ± 0.17); while Zn (0.37 ± 0.05), Ba (0.29 ± 0.04), Na (0.28 ± 0.04), Ti (0.24 ± 0.03), B (0.21 ± 0.03), F (0.2 ± 0.03), P (0.18 ± 0.02), Ga (0.16 ± 0.02), Ag(0.14 ± 0.02), Zr (0.12 ± 0.01), Mn (0.12 ± 0.01), Cu (0.11 ± 0.01), and Pd (0.10 ± 0.01) were present in trace amounts.

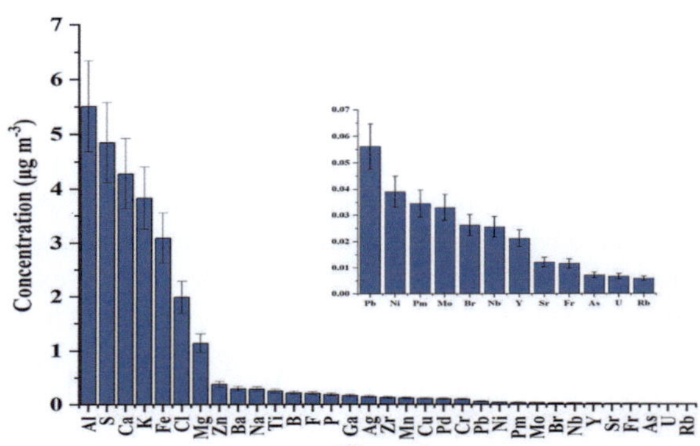

Fig. 3 Elemental composition of PM_{10}

Previous studies have reported a similar order of major elements (Ca, Na, S, Cl, Al, K, and Mg) in PM_{10} mass load over Delhi [27, 28].

3.3 Potential Source Profiling

The results of elemental concentrations obtained through WD-XRF analysis were evaluated using PCA, which successfully identified the sources of the elements. The PCA was applied to 20 quantified elements (Na, Mg, Al, Si, P, S, Cl, K, Ca, Ti, Cr, Mn, Fe, Ni, Cu, Zn, Ga, Zr, Mo, and Pb) in PM_{10} to facilitate source identification, and six distinct sources were identified (Fig. 4). The factor profile revealed that PC1 exhibited high loadings of marker elements associated with crustal, soil, and road dust, including Al, Si, Mg, Ca, Fe, and K, accounting for 27.8% of the total PM_{10} factor loading. Elements such as Al, Si, Fe, Ca, Ti, and Sr are frequently used as tracers to investigate and identify sources of dust-related emissions [29–31]. PC2 was identified as vehicular emissions showing significant loadings of Zn and Ni, which are used as a tracer of non-exhaust traffic emissions, i.e., brake and tire wears [32]. PC2 contributed 14.8% to the total identified elemental mass concentration. PC3 was identified as an industrial emission, with a contribution of 34.8%, and a considerably high loading of Zr and Mn, primarly originating from smelting and steel industries [15] PC4 resolved as combustion with higher loadings of Ti and Cr having contribution of 7.8%. PC5 had higher loading of Na and P, indicating the agricultural activity sources with contribution of 4.2%. PC6 had higher loading of S, Cl, and K, which are marker elements associated with biomass burning, with contribution of 10.4% in total factor loading.

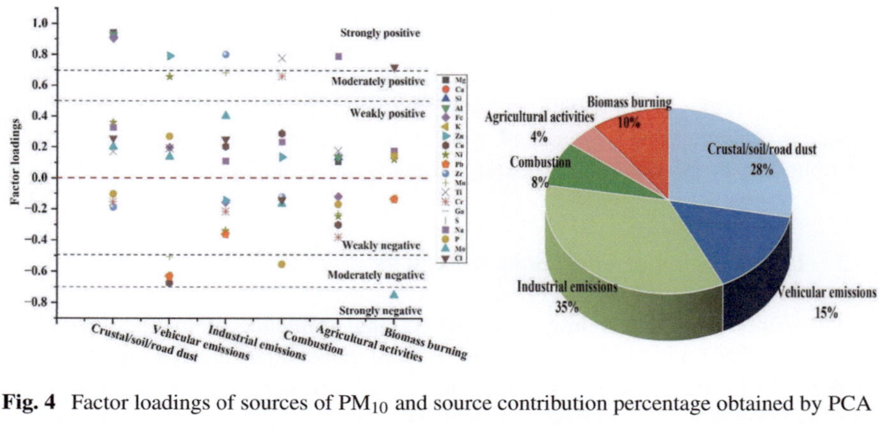

Fig. 4 Factor loadings of sources of PM_{10} and source contribution percentage obtained by PCA

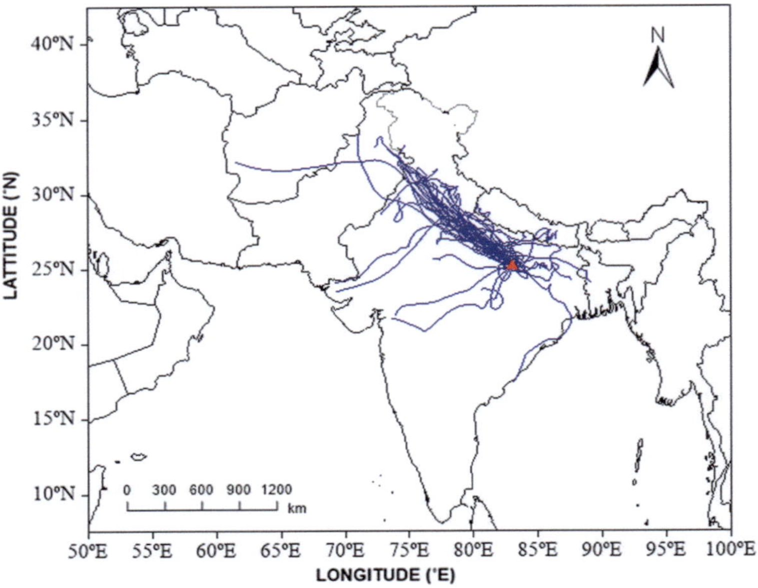

Fig. 5 72 h air-mass backward Trajectories at 500 m AGL

3.4 Source Region Mapping and Analysis

3.4.1 Air Mass Backward Trajectory

To comprehend the backward trajectories of particulate air masses reaching the sampling site in BHU, Varanasi, annual trajectories at 500 m AGL were generated (Fig. 5). Trajectory analysis revealed both local and transboundary origins of PM_{10} over the sampling site and indicated local contributions from Uttar Pradesh and regional contributions from Punjab, Haryana, Delhi, partially from Rajasthan, Gujarat, West Bengal, and some parts of Madhya Pradesh, with additional transboundary emissions from Pakistan, Afghanistan, and Nepal.

3.4.2 Identification of Local Source Regions

To have a better understanding of the directional influence of local sources, we performed CBPF with >75th percentile of mass concentration (228 μg m^{-3}) and the WS (~0–11 m s^{-1}). CBPF for Varanasi directs the source of the particulate air parcel coming from North–East, South, West, and North–West directions (Fig. 6) with major contributors including Ganga ghats, religious places, market areas, and traffic activities.

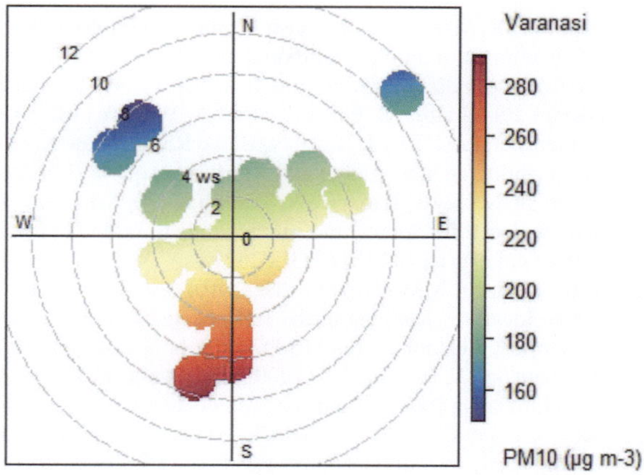

Fig. 6 Conditional Bivariate probability function (CBPF) plot of PM_{10} at study site

4 Conclusion

The study focused on identifying PM_{10} sources in the central IGP region, specifically at BHU, Varanasi, during 2019. The annual average PM_{10} concentration was $228 \pm 34\ \mu g\ m^{-3}$, with the highest levels in April. Elemental analysis showed Si, Al, S, Ca, and K as predominant, while trace elements included Zn, Na, and Cu. PCA/APCS revealed key contributors: road dust (28%), industrial emissions (35%), biomass burning (15%), and vehicular emissions (10%). Trajectory analysis identified local, regional, and transboundary contributions, while CBPF highlighted emissions from various directions, emphasizing the need for targeted mitigation strategies.

References

1. Charlson RJ, Schwartz SE, Hales JM, Cess RD, Coakley JJ, Hansen JE, Hofmann DJ (1992) Climate forcing by anthropogenic aerosols. Science 255(5043):423–430
2. Hansen J, Sato M, Ruedy R (1997) Radiative forcing and climate response. J Geophys Res Atmos (1–10) 102(D6):6831–6864 (1997)
3. Lawrence MG, Lelieveld J (2010) Atmospheric pollutant outflow from southern Asia: a review. Atmos Chem Phys Discuss 10(4):11017–11096
4. Saud T, Saxena M, Singh DP, Saraswati, Manisha Dahiya, SK Sharma, A. Datta, Ranu Gadi, TK Mandal (2013) Chemical composition of aerosol particles from direct emissions of biomass fuel used in Indo Gangetic plain. Atmos Environ 71:158–169
5. Menon S, Hansen J, Nazarenko L, Luo Y (2002) Climate effects of black carbon aerosols in China and India. Science 297(5590):2250–2253
6. Prasad MVSN, Sharma C, Arya BC, Mandal TK, Singh S, Kulshrestha MJ, Agnihotri R, Mishra SK, Sharma SK (2013) Experimental facilities to monitor various types of atmospheric parameters in the radio and atmospheric sciences division (RASD) of CSIR-national physical laboratory. Mapan-J Metrol Soc India 28(3):193–203

7. Sharma SK, Mandal TK, Rohtash, Kumar M, Gupta NC, Pathak H, Harit RC, Saxena M (2014a) Measurement of ambient ammonia over the National Capital Region of Delhi, India. Mapan-J Metrol Soc India. https://doi.org/10.1007/s12647-014-0098-9
8. Monks PS, Granier C, Fuzzi S, Stohl A, Williams ML, Akimoto H, Amann M, Baklanov A, Ballensperger U, Bey I, Blake N, Blake RS, Carslaw K, Cooper OR et al (2009) Atmospheric composition change–global and regional air quality. Atmos Environ 43(33):5268–5350
9. Chinnam N, Dey S, Tripathi SN, Sharma M (2006) Dust events in Kanpur, northern India: chemical evidence for source and implications to radiative forcing. Geophys Res Lett 33:L08803. https://doi.org/10.1029/2005GL025278
10. Ram K, Sarin MM, Tripathi SN (2010) A 1 year record of carbonaceous aerosols from an urban location (Kanpur) in the Indo Gangetic Plain: characterization, sources and temporal variability. J Geophys Res 115:D24313. https://doi.org/10.1029/2010JD014188
11. Ramanathan V, Ramana MV (2005) Persistent, widespread, and strongly absorbing haze over the Himalayan foothills and the Indo Gangetic plains. Pure Appl Geophys 162:1609–1626
12. Dey S, Tripathi SN, Mishra SK (2008) Probable mixing state of aerosols in the Indo-Gangetic Basin, northern India. Geophys Res Lett 35:L03808. https://doi.org/10.1029/2007GL032622
13. Ram K, Sarin MM (2010) Spatio-temporal variability in atmospheric abundances of EC, OC and WSOC over northern India. J Aerosol Sci 41(1):88–98
14. Jain S, Sharma SK, Srivastava MK, Chatterjee A, Vijayan N, Tripathy SS, Sharma C (2021) Chemical characterization, source apportionment and transport pathways of $PM_{2.5}$ and PM_{10} over Indo Gangetic Plain of India. Urban Climate 36:100805
15. Gupta S, Sharma SK, Tiwari P, Vijayan N (2024c) Insight Study of Trace Elements in $PM_{2.5}$ During Nine Years in Delhi, India: seasonal variation, source apportionment, and health risks assessment. Arch Environ Contam Toxicol 1–17
16. Gupta S et al (2024d) Mineralogical characteristics and sources of coarse mode particulate matter in Central Himalayas. Aerosol Sci Eng 8:121–132
17. Gupta S et al (2024b) Source apportionment and health risk assessment of PM_{10} bound carbonaceous and elemental species in the central Himalayas. Curr Nat Sci Eng 1(2):15–26
18. Sharma SK, Choudhary N, Srivastava P, Naja M, Vijayan N, Kotnala G, Mandal TK (2020) Variation of carbonaceous species and trace elements in PM_{10} at a mountain site in the central Himalayan region of India. J Atmos Chem 77:1–14
19. Song Y, Xie S, Zhang Y, Zeng L, Salmon LG, Zheng M (2006) Source apportionment of $PM_{2.5}$ in Beijing using principal component analysis/absolute principal component scores and UNMIX. Sci Total Environ 372(1):278–286
20. Thurston GD, Spengler JD (1985) A quantitative assessment of source contributions to inhalable particulate matter pollution in metropolitan Boston. Atmos Environ (1967) 19(1):9–25
21. Henry RC, Hidy GM (1979) Multivariate analysis of particulate sulfate and other air quality variables by principal components—I. Annual data from Los Angeles and New York. Atmos Environ 13:1581–1596
22. Gupta S et al (2024a) Identification of sources of coarse mode aerosol particles (PM_{10}) using ATR-FTIR and SEM-EDX spectroscopy over the Himalayan Region of India. Environ Sci Pollut Res 31:15788–15808
23. Uria-Tellaetxe I, Carslaw DC (2014) Conditional bivariate probability function for source identification. Environ Model Softw 59:1–9
24. Choudhary N, Rai A, Kuniyal JC, Srivastava P, Lata R, Dutta M, Ghosh A, Dey S, Sarkar S, Gupta S, Chaudhary S, Thakur I, Bawari A, Naja M, Vijayan N, Chatterjee A, Mandal TK, Sharma SK, Kotnala RK (2023) Chemical characterization and source apportionment of PM_{10} using receptor models over the Himalayan Region of India. Atmosphere 14:880
25. Sah D, Verma PK, Kumari KM, Lakhani A (2022) Characterisation, sources and health risk of heavy metals in $PM_{2.5}$ in Agra, India. Expo Health 49:10147
26. Banoo R, Gupta S, Gadi R, Dawar A, Vijayan N, Mandal TK, Sharma SK (2024) Chemical characteristics, morphology and source apportionment of PM_{10} over National Capital Region (NCR) of India. Environ Monit Assess 196(2):163

27. Sharma SK, Mandal TK, Saxena M, Sharma A, Gautam R (2014b) Source apportionment of PM_{10} by using positive matrix factorization of urban site in Delhi, India. Urban Clim 10:656–670
28. Jain S, Sharma SK, Mandal TK, Saxena M (2018) Source apportionment of PM_{10} in Delhi, India using PCA/APCS, UNMIX, and PMF. Particuology 37:107–118
29. Pant P, Harrison RM (2012) Critical review of receptor modeling for particulate matter: a case study of India. Atmos Environ 49:1–12
30. Sharma SK, Mandal TK, Jain S, Saraswati, Sharma A, Saxena M (2016) Source apportionment of $PM_{2.5}$ in Delhi, India using PMF model. Bull Environ Contam Toxicol 97:286–293
31. Rai P, Furger M, El Haddad I, Kumar V, Wang L, Singh A, Dixit K, Bhattu D, Petit JE, Ganguly D, Rastogi N (2020) Real-time measurement and source apportionment of elements in Delhi's atmosphere. Sci Total Environ 742:140332
32. Jain S, Sharma SK, Vijayan N, Mandal TK (2020) Seasonal characteristics of aerosols ($PM_{2.5}$ and PM_{10}) and their source apportionment using PMF: a four-year study over Delhi, India. Environ Pollut

A Review on Microplastics in Indoor Environments: Techniques for Measurement and Environmental Implications

Prerna Singh, Anuradha Singh, and Manoj Kumar

Abstract Microplastic pollution has become a worldwide environmental concern with increasing attention given to its prevalence in indoor environments. The present review studied the occurrence and sources of microplastics in indoor settings across various regions of the world, emphasizing the potential implications for human well-being and the environment. In earlier studies, various sampling, pre-treatment or extraction and analysis techniques were employed to detect, characterize and quantify microplastics particles in indoor air and dust samples. These studies revealed the widespread distribution of microplastics with numerous polymer types and sizes identified. Additionally, studies have shown that microplastics are frequently found in indoor spaces, originating from multiple sources including personal care products, cosmetics, building materials, synthetic fabrics, carpets, furniture, paint and electronic devices. Furthermore, outdoor air can carry microplastics indoors, especially in urban areas where plastic particles are abundantly present. Microplastics in indoor settings may pose a significant risk to human health; therefore, it is imperative to understand the pathways and sources of microplastics for developing effective mitigation strategies. The review also draws attention to a critical research gap in studies on indoor microplastics in India, a major contributor to pollution and plastic waste. This identifies the urgent need for conducting further studies in this region. To minimize the long-term effects on human health and the environment, scientists, policymakers and the general public should work together to discover sustainable solutions to the implications of indoor microplastics contamination.

Keywords Environment · Human health · Indoor air · Microplastics · Plastics

---

P. Singh · A. Singh · M. Kumar (✉)
Environmental Sciences & Biomedical Metrology Division, CSIR-National Physical Laboratory, New Delhi, India
e-mail: manojkumar.nplindia@csir.res.in

Academy of Scientific and Innovative Research (AcSIR), Ghaziabad, Uttar Pradesh, India

S. Yadav et al. (eds.), *Advances in Metrology*, Lecture Notes in Mechanical Engineering,
https://doi.org/10.1007/978-981-96-6418-4_4

1 Introduction

Plastics have become essential materials nowadays due to their cost-effectiveness, accessibility and versatility [1]. However, their ubiquitous presence has resulted in a new era of environmental challenge, particularly concerning microplastics (MPs). MPs are defined as plastic particles ranging from 1 μm to 5 mm in size. MPs have gained significant attention due to their widespread presence, potential ecological and human health impacts [1–3]. Understanding the distribution of sources and behaviour of MPs, particularly within indoor environments, is crucial for explaining their intricate dynamics and mitigating associated risks.

The generation of MPs comes from both primary and secondary sources. Primary sources of MPs originate from direct industrial release, while secondary MPs result from the fragmentation of bigger plastic items due to environmental forces such as UV radiation, mechanical abrasion and other factors [2]. Condition is further exacerbated by weathering processes, which change the chemical and physical characteristics of MPs and increase their environmental persistence [4].

Various sources (Fig. 1) contribute to the influx of MPs into indoor and outdoor settings. Synthetic fibres from textiles, microbeads in personal care products, tyre abrasion, sewage effluent, urban runoff, atmospheric deposition and industrial activities all contribute to the MPs in indoor and outdoor environments [5–9]. Understanding the pathways through which these sources introduce MPs into the environment is essential for comprehensively assessing exposure risks and implementing effective mitigation strategies.

The present review comprehended an examination of indoor MPs, systematically categorizing data about different dust types found within indoor environments, i.e. settled dust, fallout dust and suspended dust. Dust, comprised of a heterogeneous mixture of organic and inorganic matter, functions as an effective carrier and reservoir for MPs. This dynamic interaction facilitates the widespread dispersal and subsequent deposition of MPs within diverse environmental matrices. Through diverse measurement techniques, the earlier studies identified major polymers, as well as the shapes and sizes of MPs present within the collected indoor dust samples. Furthermore, the review highlighted a significant research gap in the identification of MPs within dust samples, particularly within the geographic regions of India. This is especially relevant given India's status as one of the leading contributors to plastic waste generation and pollution globally. Additionally, the lack of studies in this domain within the Indian context emphasizes the need for comprehensive scientific investigations to identify the prevalence, distribution and potential environmental implications of MPs in dust samples of India, thereby facilitating informed policy interventions and mitigation strategies.

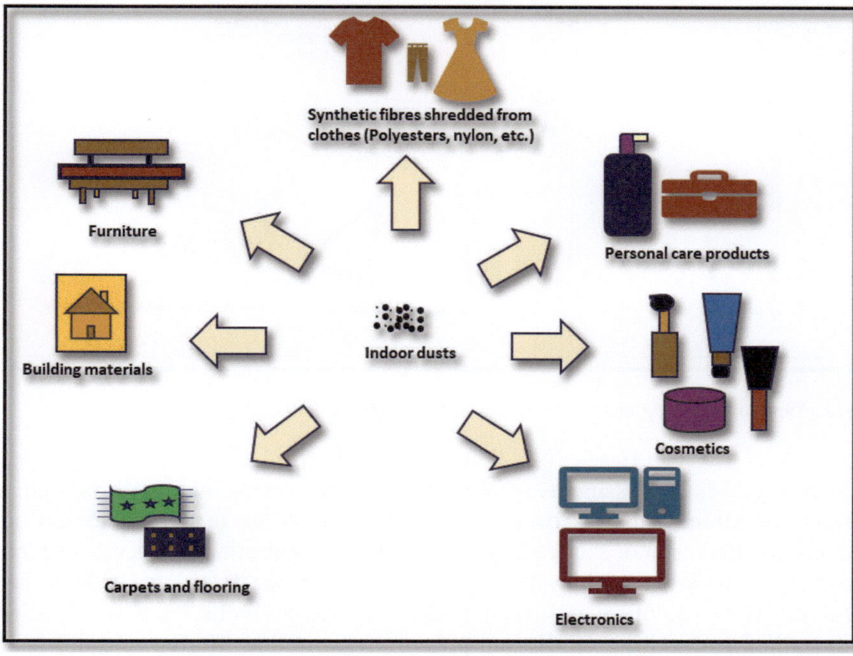

Fig. 1 Sources of indoor microplastics (MPs)

2 Techniques for Measurement of Indoor Microplastics (MPs)

Techniques for quantifying MPs in dust samples involve distinct methodologies, enabling a comprehensive evaluation of contamination levels. These methodologies typically consist of three main steps: Sampling, pre-treatment or extraction and identification. Sampling methods are categorized into active and passive techniques, each with unique advantages and challenges (Fig. 2). Passive methods, preferred for their simplicity and cost-effectiveness, require no electricity and are suitable for long-term data collection. They involve two approaches; collecting MPs from atmospheric deposition/fallout samples using devices like metallic funnels, glass bottles, stainless steel basins, glass beakers and petri dishes positioned at heights of 1.2–1.8 m, and collecting settled dust from floors using vacuum cleaners, dustpans and brushes [10–15]. In contrast, active methods involve pumping a fixed amount of air over a period using air samplers, producing more precise results for MPs in suspended dust [16, 17]. Passive sampling methodologies predominate in the majority of investigations concerning indoor MPs, as indicated by the findings of this review. Quantification of MPs in various compartments is based on the sample type, with deposited indoor dust measured in MPs number per gram of dust, atmospheric dust deposition in MPs number per square metre per day and suspended dust in MPs number per cubic metre.

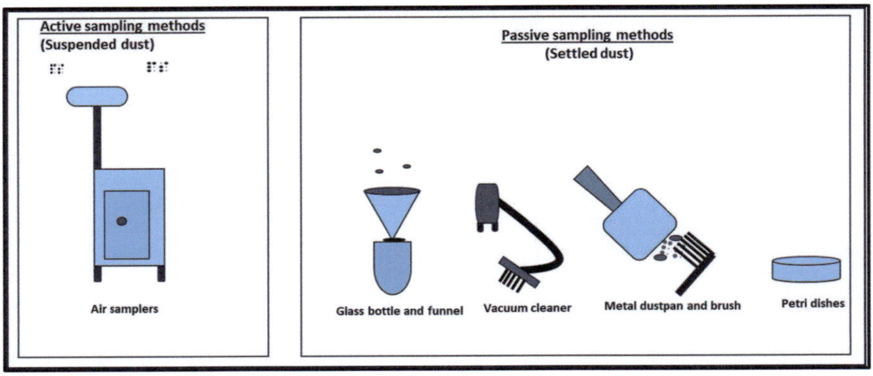

Fig. 2 Types of sampling method for indoor microplastics (MPs) collection

Isolating MPs from samples involves pre-treatment/extraction methods such as organic digestion, density separation and filtration. While many studies did not perform pre-treatment or extraction due to the low concentration of organic matter in samples [18], those that employ it typically use methods like digestion with hydrogen peroxide (H_2O_2) solution, sodium hypochlorite (NaClO) solution, base digestion with sodium hydroxide (NaOH) and potassium hydroxide (KOH), acid digestion with nitric acid (HNO_3) and sulfuric acid (H_2SO_4), and various enzymatic methods [19–23]. Density separation utilizes salt solutions of various densities like zinc chloride ($ZnCl_2$), sodium iodide (NaI) and sodium chloride (NaCl) to isolate MPs from the sample matrix [10, 15, 24]. The density of the majority of common polymers ranges from around 0.9–1.4 g/cm^3. This includes polymers like polyethylene (HDPE; density 0.96 g/cm^3) and (LDPE; density 0.92 g/cm^3), polypropylene (PP; density 0.91–0.95 g/cm^3), polyvinyl chloride (PVC; density 1.35–1.39 g/cm^3), polystyrene (PS; density 1.05 g/cm^3) and polyethylene terephthalate (PET; density 1.38–1.41 g/cm^3) [25]. Zinc chloride ($ZnCl_2$) proved useful in extracting MPs from diverse environmental compartments due to its high density. With a density of approximately (1.6–1.7 g/cm^3), $ZnCl_2$ is better than many saturated salt solutions conventionally used for the extraction of MPs, such as sodium chloride (NaCl) or sodium bromide (NaBr), in density separation. The increased density of $ZnCl_2$ facilitates the efficient separation of denser MPs polymers such as PVC and PET, which may prove challenging to isolate using lower-density solutions [26].

Analysis of MPs in the dust samples can be carried out through physical and chemical characterization methods. Physical characteristics including size, shape and colour are identified using various microscopy techniques, while chemical properties and polymer identification are determined through spectroscopy methods. Fourier transform infrared spectroscopy (FTIR) combined with microscopy is a commonly adopted approach for comprehensive characterization, considering factors such as detection limits, identification time and cost-effectiveness (Table 1). Additionally, FTIR spectroscopy plays a significant role in identifying the polymer composition of

MPs, further enhancing the comprehensive characterization of indoor MPs pollution [27, 28].

In the existing studies on MPs, FTIR and Raman's spectroscopy are the most common techniques used in analysing the composition of MPs due to their vibrational spectroscopic principles. The way light interacts with molecules to produce specific vibrations is the basis for these methods. A particular peak pattern or spectrum that is produced functions as a "molecular fingerprint" for MPs, revealing details about their actual compositions and identities. Distinctive peaks were categorized based on wavenumbers, revealing a varied range reflecting plastic complexity. The variation in FTIR and Raman spectra of MPs for the same type of polymer is due to several factors. Additives like plasticizers, stabilizers and pigments, as well as fillers, can alter spectral properties. Environmental factors such as UV radiation, temperature changes, mechanical abrasion and chemical interactions also contribute to these differences. Additionally, variations in sample preparation and measurement conditions also impact the spectra. Together, these factors influence the observed wavenumbers and spectral features. Despite the variability present, certain characteristic peaks consistently emerge in the majority of cases; therefore, establishing a valuable repository of these identifying peaks would facilitate easier recognition and analysis.

In addition, Scanning Electron Microscopy (SEM-EDX) is another method which is widely employed in many studies. It provides insights into surface characteristics, degradation patterns and chemical composition of MPs [29]. Pits, grooves, flakes and jagged edges are just a few of the surface degradation patterns that can be seen through SEM. Pits are tiny holes that show signs of erosion and wear. Linear channels that indicate abrasion and mechanical shaping are called grooves. Rough edges are jagged areas that show fractures and natural irregularities, whereas flakes are detachable pieces of material created by stress and degradation. Consistent findings regarding the chemical composition of indoor MPs have been reported across various studies. Analysis of house dust samples using EDX confirmed the prevalence of carbon, oxygen and nitrogen [15, 30, 31]. However, one study in Iran found unexpectedly high levels of mercury (3–7%), yet the authors did not explore this further [32].

3 Worldwide Studies on Indoor Microplastics (MPs)

Indoor environments are now emerging as hotspots for MPs contamination. Worldwide studies on indoor MPs have revealed a widespread presence of numerous types of MPs; synthetic particles pervade indoor spaces across diverse geographic regions, posing significant risks to both human health and environmental integrity. Findings from various studies provide insights into the prevalence and diversity of MPs in indoor environments across various geographical locations. Various studies have detected MPs in different indoor settings. The first study, conducted by Dris et al. [10], focused on apartments in Paris. Subsequent research has been carried out in

Table 1 Worldwide studies on indoor MPs

S. No	References	Study area	Sampling type of dust samples (active/passive)	Sampling devices and characterization technique used	Major findings: polymers/shapes/size-range of (mps) in indoor dust samples
1	[10]	Paris	Passive	Settled indoor dust/Vacuum cleaner Stereomicroscope, FTIR (ATR MODE)	Major polymer: PP dominant shape: fibres size range: 50–3250 μm
2	[35]	Aarhus, Denmark	Active	Suspended dust/breathing thermal manikin (BTM) device FPA-μ FTIR	Polyester was predominant polymer (81%), followed by PE (6%), NY (5%), PP (2%), Other polymers (6%) Shapes: fibres (13%), fragments (87%)
3	[36]	Thirty nine major cities of China	Passive	Settled indoor dust/Hog bristle brushes and dustpan Optical microscope, μ FTIR, LC-MS/MS	PET and PC MPs were detected in the samples Major polymer: PET Dominant shape: fibres
4	[48]	Shanghai, China	Passive	Atmospheric fallout/Stainless steel basins (Open area: 0.08 m^2) FTIR	Dominant shapes: fibres Majority MPs size range: 50–2000 μm (80% of MPs)
5	[11]	Surabaya, Indonesia	Passive	Settled indoor dust/vacuum cleaner Digital microscope	Fibres (85%), fragments and films shapes observed Dominant size range: 3000–3500 μm
6	[49]	Aveiro, Portugal	Active	Suspended dust/portable air sampler Stereomicroscope	Size range: 17–3669 μm, Median size: 250 μm

(continued)

Table 1 (continued)

S. No	References	Study area	Sampling type of dust samples (active/ passive)	Sampling devices and characterization technique used	Major findings: polymers/shapes/ size-range of (mps) in indoor dust samples
7	[33]	Surabaya, Indonesia	Passive	Atmospheric deposition/transparent container of size of 30 × 20 × 10 cm Microscope, FTIR	Polymers like PET, polyester, cellophane, chipboard and alkyd resin Only Fibres (90%) and Fragments were found Dominant size range: 1000–1500 μm
8	[14]	Twelve different countries	Passive	Settled indoor dust/ vacuum cleaner or by sweeping with nylon brush in a dust pan HPLC-MS/MS	PET and PC polymers identified
9	[13]	Hull and Humber region, UK	Passive	Atmospheric deposition/1 L glass beakers (Area: 0.0095 m^2) Stereomicroscope, μ FTIR	PET (63%), PA (6%), PP (4%), Acrylic (4%), PAN (3%) Dominant shape: Fibres (90%) Dominant size range: (5–250 μm)
10	[12]	Sydney, Australia	Passive	Atmospheric deposition/glass petri dishes (Diameter: 12 cm) Stereomicroscope, Fluorescence microscope, μ FTIR	PE (25%), polyester (17%) including PET, Polyamide (16%), Polyvinyl (15%) Polymers were found Fibres (99%) Size range: 50–600 μ m

(continued)

Table 1 (continued)

S. No	References	Study area	Sampling type of dust samples (active/passive)	Sampling devices and characterization technique used	Major findings: polymers/shapes/size-range of (mps) in indoor dust samples
11	[16]	Wenzhou, Eastern China	Active	Suspended dust/total suspended particulate sampler Microscope μ FTIR	Polyester (28.4%), PA (20.54%), PP (16.3%) polymers found Fibres and fragments observed; predominant shape: fragment 5–30 μm size range; 54.1–65.2% 30–100 μm size range; 25.3–32.8% MP larger than 100 μm; 11%
12	[17]	Aveiro, Portugal	Active	Suspended dust/Portable air sampler Microscope (Nile Red)	Size of fibres were in the range of 37 to 10822 μm, with a median of 445 μm Shapes: Fragments (71.4%), spheres (28.6%)
13	[32]	Iran	Passive	Settled dust/Horsetail strands brush and a steel dustpan Optical microscope, SEM-EDX, Raman	Polymers were PET (50%) > PP (35%) > PS (10%) > NY (5%) Sheets and fragments were observed Dominant size range: 500–1000 μm
14	[34]	Shanghai, China	Active	Suspended dust/air samplers Raman	PE (73.82%), Polyester (9.21%), Phenolic resin (9.05), PVC (3.06%), Cotton (1.36%), PP (0.57%), PU (0.34%), Rubber (0.18%) Predominate shape: Fragments (85%) Size range: 2.40–2181.48 μm

(continued)

Table 1 (continued)

S. No	References	Study area	Sampling type of dust samples (active/passive)	Sampling devices and characterization technique used	Major findings: polymers/shapes/size-range of (mps) in indoor dust samples
15	[28]	Sri Lanka	Active	Suspended dust/air sampler Stereomicroscope, μ FTIR	PET (60%), followed by polyester PES (14%), acrylic (12%), polyamide PA (6%), polypropylene PP (6%) and polystyrene PS (2%) Shape: Fibres (98%), fragments Size range: 50–5000 μm
16	[15]	Bushehr port and Shiraz, Iran	Passive	Settled dust/Horsetail brush and a steel dustpan Microscope, μ Raman, SEM-EDX	PE (42%), PC (28%), PP (18.6%), PET (10%), Nylon (1.4%) Shapes: Fibres (85%), fragment (13%), film (2%) Size most abundant > 1000 μm
17	[31]	New Jersey, United States	Passive	Atmospheric deposition/Whatman quartz filter (2.2 μm, diameter: 47 mm) in a glass Petri dish Microscope, Raman, SEM	Polymers: (PS), (PET), (PE), (PVC) and (PP) polymers identified Shapes: Fibres, films and fragments
18	[30]	Shiraz, Iran	Passive	Settled dust/Horse-hair brush Raman, SEM-EDX, Binocular microscope	Polymers mainly constructed of PET or PP MPs were dominated by Fibres (>250) μm in length Shape: fibres (85%), Fragments (4%)
19	[37]	Mumbai, India	Passive	Atmospheric deposition/ceramic mug Digital microscope	Fragments

schools and offices [11, 33], living rooms [34] and a wide range of other indoor environments such as hospitals, mosques, kindergartens and universities [15]. In Paris, France [10], settled indoor dust collected via vacuum cleaner revealed predominantly polypropylene (PP) MPs in the form of fibres ranging from 50 to 3250 μm. In Aarhus, Denmark, Vianello et al. [35], showed polyester (81%) as the predominant polymer in suspended dust, with fragments (87%) and fibres (13%) being the dominant shapes. Surabaya, Indonesia [11, 33], exhibited a high prevalence of fibres, fragments and films in indoor dust, with dominant sizes ranging from (3000–3500) to (1000–1500) μm. The Hull and Humber region of UK [13] found predominantly polyethylene terephthalate (PET) (63%) fibres in atmospheric deposition, while Sydney, Australia [12], identified majority of the fibres primarily composed of polyethylene (PE), polyester (PES), polyamide (PA), polyvinylchloride (PVC), polystyrene (PS) and polyacrylic (PA) types of polymers. Shanghai, China [34], showed predominantly polyethylene (PE) (73.82%) and fragmented MPs (85%) in suspended dust. In a global investigation of MPs found in household dust across twelve diverse countries, including China, Greece, Colombia, Kuwait, Japan and Saudi Arabia, MPs were identified in each one of the 286 dust samples. The concentrations ranged from 38,000 to 120,000 $\mu g\ g^{-1}$ (with a median of 5,900 $\mu g\ g^{-1}$) [14]. Liu et al. [36] conducted a study analysing indoor and outdoor samples of dust from 39 cities across China to assess two types of MPs, specifically polycarbonate (PC) and polyethylene terephthalate (PET). The findings revealed the presence of PET MPs in every sample, with notable concentrations ranging from 1550 to 120,000 mg kg^{-1} indoors and 212–9020 mg kg^{-1} outdoors. Polycarbonate MPs were identified in 70% of the samples, with a median concentrations of 2.0 mg kg^{-1} outdoors and 4.6 mg kg^{-1} indoors [36]. These studies show how indoor MPs are distributed worldwide and emphasize how different polymers, forms and sizes are common in indoor environments. To minimize MPs pollution indoors, urgent mitigation measures are required including extensive waste management procedures, product design innovations and public awareness initiatives. Addressing this complex challenge calls for cooperative action at the local, national and international levels to protect human health and the integrity of the ecosystem.

Table 1 summarizes findings from various studies on MPs pollution in different regions across the world, focusing on the sampling type, dust sample characteristics and the identified polymers, shapes and sizes of MPs. The table reveals a global presence of MPs in diverse environments, emphasizing varied polymer compositions, shapes and sizes. Common polymers such as PET and PP dominate in the indoor environments, with fibres being the prevailing shape. Size ranges vary widely, highlighting the need for standardized assessments to comprehensively understand and address MPs pollution in different regions.

4 Studies of Indoor Microplastics (MPs) from India

Studies on indoor MPs in India have been notably less compared to outdoor investigations, reflecting a gap in understanding the extent and implications of MPs contamination within indoor environments. Among the limited indoor studies conducted, findings revealed a predominance of fibrous and fragments type of MPs, primarily originating from materials such as polyethylene (PE) and polyethylene terephthalate (PET). For instance, in Mumbai, passive sampling of atmospheric deposition in both indoor and outdoor indicated the presence of MPs fragments, although detailed characterization and abundance metrics were not provided [37]. In another study, the author conducted a study on the presence of MPs in various environmental mediums within the residential campus of IIT Bombay, focusing on drinking water, air and food samples. Results indicated that the average daily exposure to MPs was 382 ± 205 particles per person for drinking water and 594 ± 269 particles for air. Food exhibited the highest exposure with 1036 ± 493 particles per person. Mass-based exposure ranged from (122.25 ± 177.38 mg to 202.80 ± 294.25 mg) per person, primarily sourced from food. Contributions from air (61.85 ± 16.63 μg to 102.59 ± 27.59 μg) and drinking water (55.21 ± 53.86 μg to 91.6 ± 89.34 μg) were comparatively negligible. Similarly, a study in Patna by Zhang et al. [14] focused on settled dust samples collected indoors, highlighting the prevalence of PET MPs, yet specific shape and abundance data were not reported.

In the indoor and outdoor air of Ranchi in Jharkhand, India, MPs primarily appeared as fibres, followed by fragments and small number of films. Outdoor air predominantly contained polyethylene (PE), polypropylene (PP) and polystyrene (PS), while indoor air primarily composed of polyvinyl chloride (PVC), along with cellulose possibly originating from textiles and clothing. Moreover, elemental mapping confirmed the presence of diverse elements on the MPs. MPs concentrations showed significant seasonal fluctuations, with wind speed and direction impacting their abundance, indicating their potential as reservoirs for environmental contaminants [38]. Despite these insights, indoor studies remain limited in scope and geographical coverage of India.

The lack of indoor MPs research in India is concerning given the potential risks posed by indoor exposure to MPs, including inhalation, ingestion and dermal contact, which can have adverse effects on human health. Indoor MPs originate from a variety of sources, including household products, textiles and building materials, making indoor environments reservoirs for MPs accumulation (Fig. 1). Furthermore, factors such as inadequate ventilation, high population density and various anthropogenic activities contribute to elevated indoor MPs levels in India. Given the country's rapid urbanization, industrialization and growing consumption of plastic products, there is a crucial need for more comprehensive and systematic studies on indoor MPs. Future research should emphasize on understanding exposure pathways, assessing health risks and formulating effective mitigation strategies to safeguard public health and environmental quality in India. Increased investment in indoor MPs research, coupled with interdisciplinary collaboration and public awareness initiatives, is crucial for addressing this emerging environmental and public health challenge in the country.

5 Sources and Pathway of Indoor MPs

The pathways by which MPs enter indoor environments are complex, involving a range of factors such as atmospheric deposition, water and food contamination, and anthropogenic actions, potentially including respiratory inhalation (airborne pathway), oral ingestion (via food, drinking water and dust) and dermal contact (with dust, textiles and personal care items) [29].

Particles can infiltrate indoor environments through various pathways, including infiltration, tracking and penetration. Infiltration means the direct exchange of air via open apertures like windows, doors and through mechanical ventilation systems lacking particle filtration. Tracking involves introducing particles, such as those carried by footwear and clothing. Penetration occurs when particles enter through minute cracks in the building envelope [39].

Airborne MPs generated by various processes settle onto surfaces or inhaled by occupants, posing potential health risks. On deposition within indoor environments, MPs undergo degradation, aggregation and resuspension, exerting significant influence over their distribution patterns and subsequent exposure pathways. Water contamination, either through tap water or plumbing systems, is another significant pathway for MPs. Additionally, contaminated food products can introduce MPs into indoor environments during preparation or consumption. Human activities such as cleaning and agitation of materials further resuspend MPs, perpetuating their presence indoors. Indoor-specific sources like shedding microfibres from textiles and synthetic furnishings contribute to this contamination. Therefore, formulating effective strategies to decrease indoor MPs pollution and protect human health and environmental integrity requires a thorough understanding of these varied sources and pathways [40].

6 Environmental and Health Implications

The presence of indoor MPs can have profound impacts on both the environment and the health of people residing indoors. These tiny plastic particles, originating from sources like synthetic textiles, personal care products, and many more, can become airborne or contaminate the indoor environment, leading to respiratory issues and potential ingestion through drinking water and food. When indoor MPs are released into the environment, they affect the ecosystem by causing ecological disruption, bioaccumulating in species and magnifying via the food chain, all of which can be detrimental to human health. Countries with inadequate waste management exacerbate the problem, as MPs persist in landfills and leach into soil and groundwater, further contaminating the environment. Furthermore, MPs also act as vectors for the transportation of pollutants and microbial pathogens, thereby presenting potential hazards to human health through routes of exposure including inhalation, ingestion and dermal contact [41–43].

With mean concentration of 9.8 MPs particles per cubic metre, the findings of Cox et al. [44] indicated that adult males inhale approximately 170 MPs particles per day, while adult females inhale 132 MPs particles per day. Children, on the other hand, were exposed to around 110 particles per day for males and 97 for females. These estimates represent approximately half or more of the total daily exposure via all pathways in the studied population groups, highlighting inhalation as the primary route of human exposure to MPs. Furthermore, Vianello et al. [35] documented that, on average, male adults involve in light activity inhale an estimated mean of 272 MPs particles per day. The elevated risk induced by plastic additives such as phthalates, bisphenols, organophosphates and biocides, along with the persistence and bio-accumulative traits observed due to substances like brominated flame retardants and chlorinated paraffin, is intensified by the substantial surface area to mass ratios of MPs. Additionally, there is a greater chance that chemicals from MPs will leak out of the plastic polymer when humans are exposed to MPs [45–47]. A recent study on MPs in indoor dust from thirty-two Australian homes found children ingested an average of 6.1 mg per kilogram of body weight per year (equivalent to an estimated daily intake of 16,712 nanograms per kilogram of body weight per day), while adults ingested 0.5 mg per kilogram of body weight per year (equivalent to an estimated daily intake of 1370 nanograms per kilogram of body weight per day) [12]. Addressing these implications requires comprehensive strategies such as reducing indoor plastic usage, improving air and water quality management, implementing effective waste disposal practices, and promoting sustainable alternatives. Further research is necessary to understand the long-term effects and to develop practical mitigation strategies that will protect the ecosystem and public health from the detrimental impacts of indoor MPs.

7 Conclusion

The growing concern of MPs pollution in indoor settings needs comprehensive understanding and immediate action to mitigate its harmful effects on human health and the environment. Indoor surroundings, earlier considered safe from outdoor pollutants, are now identified as hotspots for MPs contamination. This review has presented the occurrence, sources, techniques of measurement and environmental effects of MPs in indoor settings, drawing attention to the significant research gaps in research, particularly in developing regions like India. Studies worldwide have revealed the widespread distribution of MPs in indoor spaces, originating from several sources such as personal care products, cosmetics, building materials, synthetic fabrics like polyester, nylon, and acrylic clothes, carpets and flooring, furniture, paint, electronic devices, etc. Techniques for measuring MPs in indoor dust samples involve various methodologies, including sampling, pre-treatment or extraction and identification. Passive sampling methods predominate in indoor MPs studies due to their simplicity and cost-effectiveness. Chemical and physical characterization methods play crucial roles in identifying the polymer composition and properties of MPs. The review also highlights the necessity for a repository of characteristic peaks in the Raman and FTIR spectra. This repository would facilitate the identification process by emphasizing common molecular vibrations, despite the variability arising from the complexity of plastic materials. The inclusion of these significant peaks will enhance future research efforts, improve our understanding of the composition of MPs, and serve as a valuable resource for subsequent characterization and quality assessment. Furthermore, this review also emphasizes the urgent need for more extensive research on indoor MPs, particularly in India, considering its significant contribution to pollution and plastic waste generation. Despite limited studies, findings showed the presence of MPs in indoor settings across the country, raising concerns for both; human health and environmental well-being. The pathways through which MPs enter indoor environments are complex and multifaceted, involving air dispersion, water and food contamination and anthropogenic actions. Once indoors, MPs pose risks through respiratory inhalation, oral ingestion and dermal contact, with potential health impacts intensified by the presence of harmful chemicals and additives associated with plastic pollution. Scientists, decision-makers, and the general public must work together to address the environmental and health effects of indoor MPs. Future strategies should include reducing plastic consumption, enhancing waste management procedures and promoting sustainable alternatives. Continued research is imperative to know the long-term effects and develop effective mitigation measures, ensuring the protection of both human well-being and the environmental quality in indoor spaces.

Acknowledgements The authors are grateful to the Director of CSIR-National Physical Laboratory for providing the infrastructure and support for this study. Prerna Singh (JRF) and Anuradha (JRF) would like to acknowledge the University Grant Commission (UGC), Govt. of India for providing research grants and acknowledge the support of the Academy of Scientific and Innovative Research (AcSIR) in facilitating her Ph.D. studies.

References

1. GESAMP (Group of Experts on the Scientific Aspects of Marine Environmental Protection) (2015) Sources, fate and effects of microplastics in the marine environment: a global assessment. http://www.gesamp.org/site/assets/files/1275/sources-fate-and-effects-of-microplastics-in-the-marine-environment-part-2-of-a-global-assessment-en.pdf. Accessed 01 June 2024
2. Crawford CB, Quinn B (2017) Plastic production, waste and legislation. Microplastic Pollut 30:39–56
3. Thompson RC, Olsen Y, Mitchell RP, Davis A, Rowland SJ, John AW, Russell AE (2004) Lost at sea: where is all the plastic? Science 304(5672):838–838
4. Pinlova B, Nowack B (2024) From cracks to secondary microplastics-surface characterization of polyethylene terephthalate (PET) during weathering. Chemosphere 352:141305
5. Henry B, Laitala K, Klepp IG (2019) Microfibres from apparel and home textiles: prospects for including microplastics in environmental sustainability assessment. Sci Total Environ 652:483–494
6. Sommer F, Dietze V, Baum A, Sauer J, Gilge S, Maschowski C, Gieré R (2018) Tire abrasion as a major source of microplastics in the environment. Aerosol Air Qual Res 18(8):2014–2028
7. Gatidou G, Arvaniti OS, Stasinakis AS (2019) Review on the occurrence and fate of microplastics in Sewage treatment plants. J Hazard Mater 367:504–512
8. Rezania S, Park J, Din MFM, Taib SM, Talaiekhozani A, Yadav KK, Kamyab H (2018) Microplastics pollution in different aquatic environments and biota: a review of recent studies. Mar Pollut Bull 133:191–208
9. Dris R, Gasperi J, Rocher V, Saad M, Renault N, Tassin B (2015) Microplastic contamination in an urban area: a case study in Greater Paris. Environ Chem 12(5):592–599

10. Dris R, Gasperi J, Mirande C, Mandin C, Guerrouache M, Langlois V, Tassin B (2017) A first overview of textile fibers, including microplastics, in indoor and outdoor environments. Environ Pollut 221:453–458
11. Bahrina I, Syafei AD, Satoto R, Jiang JJ, Nurasrin NR, Assomadi AF, Nasir M (2020) An occupant-based overview of microplastics in indoor environments in the City of Surabaya, Indonesia. J Ecol Eng 21(8):236–242
12. Soltani NS, Taylor MP, Wilson SP (2021) Quantification and exposure assessment of microplastics in Australian indoor house dust. Environ Pollut 283:117064
13. Jenner LC, Sadofsky LR, Danopoulos E, Rotchell JM (2021) Household indoor microplastics within the Humber region (United Kingdom): quantification and chemical characterisation of particles present. Atmos Environ 259:118512
14. Zhang J, Wang L, Kannan K (2020) Microplastics in house dust from 12 countries and associated human exposure. Environ Int 134:105314
15. Kashfi FS, Ramavandi B, Arfaeinia H, Mohammadi A, Saeedi R, De-la-Torre GE, Dobaradaran S (2022) Occurrence and exposure assessment of microplastics in indoor dusts of buildings with different applications in Bushehr and Shiraz cities, Iran. Sci Total Environ 829:154651
16. Liao Z, Ji X, Ma Y, Lv B, Huang W, Zhu X, Shang X (2021) Airborne microplastics in indoor and outdoor environments of a coastal city in Eastern China. J Hazard Mater 417:126007
17. Xumiao L, Prata JC, Alves JR, Duarte AC, Rocha-Santos T, Cerqueira M (2021) Airborne microplastics and fibers in indoor residential environments in Aveiro, Portugal. Environ Adv 6:100134
18. Beaurepaire M, Dris R, Gasperi J, Tassin B (2021) Microplastics in the atmospheric compartment: a comprehensive review on methods, results on their occurrence and determining factors. Curr Opin Food Sci 41:159–168
19. Allen S, Allen D, Phoenix VR, Le Roux G, Durántez Jiménez P, Simonneau A, Galop D (2019) Atmospheric transport and deposition of microplastics in a remote mountain catchment. Nat Geosci 12(5):339–344
20. Uddin S, Fowler SW, Saeed T (2020) Microplastic particles in the Persian/Arabian Gulf–a review on sampling and identification. Mar Pollut Bull 154:111100
21. Lödger MG, Imhof HK, Ladehoff M, Löschel LA, Lorenz C, Mintenig S, Gerdts G (2017) Enzymatic purification of microplastics in environmental samples. Environ Sci Technol 51(24):14283–14292
22. Dris R, Gasperi J, Saad M, Mirande C, Tassin B (2016) Synthetic fibers in atmospheric fallout: a source of microplastics in the environment? Mar Pollut Bull 104(1–2):290–293
23. Renner G, Schmidt TC, Schram J (2018) Analytical methodologies for monitoring micro (nano) plastics: which are fit for purpose? Curr Opin Environ Sci Health 1:55–61
24. Patchaiyappan A, Dowarah K, Ahmed SZ, Prabakaran M, Jayakumar S, Thirunavukkarasu C, Devipriya SP (2021) Prevalence and characteristics of microplastics present in the street dust collected from Chennai metropolitan city, India. Chemosphere 269:128757
25. Ruggero F, Gori R, Lubello C (2020) Methodologies for microplastics recovery and identification in heterogeneous solid matrices: a review. J Polym Environ 28:739–748
26. Radford F, Zapata-Restrepo LM, Horton AA, Hudson MD, Shaw PJ, Williams ID (2021) Developing a systematic method for extraction of microplastics in soils. Anal Methods 13(14):1695–1705
27. Shim WJ, Hong SH, Eo SE (2017) Identification methods in microplastic analysis: a review. Anal Methods 9:1384–1391
28. Perera K, Ziajahromi S, Bengtson Nash S, Manage PM, Leusch FD (2022) Airborne microplastics in indoor and outdoor environments of a developing country in south Asia: abundance, distribution, morphology, and possible sources. Environ Sci Technol 56(23):16676–16685
29. Ageel HK, Harrad S, Abdallah MAE (2022) Occurrence, human exposure, and risk of microplastics in the indoor environment. Environ Sci Process Impacts 24(1):17–31
30. Abbasi S, Turner A, Sharifi R, Nematollahi MJ, Keshavarzifard M, Moghtaderi T (2022) Microplastics in the school classrooms of Shiraz, Iran. Build Environ 207:108562

31. Yao Y, Glamoclija M, Murphy A, Gao Y (2022) Characterization of microplastics in indoor and ambient air in northern New Jersey. Environ Res 207:112142
32. Nematollahi MJ, Zarei F, Keshavarzi B, Zarei M, Moore F, Busquets R, Kelly FJ (2022) Microplastic occurrence in settled indoor dust in schools. Sci Total Environ 807:150984
33. Pratiwi A, Syafei AD, Assomadi AF, Boedisantoso R, Hermana J (2020) Microplastic characterization based on the number of occupants. AIP Conf Proc 2296(1)
34. Xie Y, Li Y, Feng Y, Cheng W, Wang Y (2022) Inhalable microplastics prevails in air: Exploring the size detection limit. Environ Int 162:107151
35. Vianello A, Jensen RL, Liu L, Vollertsen J (2019) Simulating human exposure to indoor airborne microplastics using a Breathing Thermal Manikin. Sci Rep 9(1):8670
36. Liu C, Li J, Zhang Y, Wang L, Deng J, Gao Y, Sun H (2019) Widespread distribution of PET and PC microplastics in dust in urban China and their estimated human exposure. Environ Int 128:116–124
37. Yadav H, Sethulekshmi S, Shriwastav A (2022) Estimation of microplastic exposure via the composite sampling of drinking water, respirable air, and cooked food from Mumbai, India. Environ Res 214:113735
38. Nandi S, Kumar RN, Dhandapani A, Iqbal J (2024) Characterization of microplastics in outdoor and indoor air in Ranchi, Jharkhand, India: First insights from the region. Environ Pollut 346:123543
39. Nazaroff WW (2004) Indoor particle dynamics. Indoor Air 14:175–183
40. Sheraz M, Kim J, Kim J (2023) Nano/microplastics in indoor air: a critical review of synthesis routes for toxicity testing and preventative measure strategies. Process Saf Environ Prot 180:274–304
41. Prata JC (2018) Airborne microplastics: consequences to human health? Environ Pollut 234:115–126
42. Gasperi J, Wright SL, Dris R, Collard F, Mandin C, Guerrouache M, Tassin B (2018) Microplastics in air: are we breathing it in? Curr Opin Environ Sci Health 1:1–5
43. Han D, Currell MJ (2017) Persistent organic pollutants in China's surface water systems. Sci Total Environ 580:602–625
44. Cox KD, Covernton GA, Davies HL, Dower JF, Juanes F, Dudas SE (2019) Human consumption of microplastics. Environ Sci Technol 53(12):7068–7074
45. Wu P, Huang J, Zheng Y, Yang Y, Zhang Y, He F, Gao B (2019) Environmental occurrences, fate, and impacts of microplastics. Ecotoxicol Environ Saf 184:109612
46. Sapea A (2019) Scientific perspective on microplastics in nature and society. Science Advice for Policy by European Academies, Berlin
47. Jiang B, Kauffman AE, Li L, McFee W, Cai B, Weinstein J, Xiao S (2020) Health impacts of environmental contamination of micro-and nanoplastics: a review. Environ Health Prev Med 25:1–15
48. Zhang Q, Zhao Y, Du F, Cai H, Wang G, Shi H (2020) Microplastic fallout in different indoor environments. Environ Sci Technol 54(11):6530–6539
49. Prata JC, Castro JL, da Costa JP, Duarte AC, Rocha-Santos T, Cerqueira M (2020) The importance of contamination control in airborne fibers and microplastic sampling: experiences from indoor and outdoor air sampling in Aveiro, Portugal. Mar Pollut Bull 159:111522

Evaluating the Stratum 1 NTP Server Performance at CSIR-NPL for Nation-Wide Time Synchronization

Divya Singh Yadav, Deepak Sharma, Preeti Kandpal, and Ashish Agarwal

Abstract CSIR-NPL is the National Metrology Institute (NMI) of India by an act of parliament. It holds the responsibility to realize, maintain and disseminate the Indian Standard Time (IST), which is traceable to Coordinated Universal Time (UTC) with an accuracy of a few nanoseconds. Dissemination of IST is done via various methods, including satellite links, network time services, and calibration of clocks and oscillators. CSIR-NPL also offers services for delay calibrations of satellite timing receivers and measuring the frequency stability of oscillators. CSIR-NPL has maintained the stack of Network Time Protocol Servers (NTP servers) to disseminate IST over the network. It uses a hierarchical, semi-layered system of time sources. NTP servers at CSIR-NPL are designated as stratum 1 NTP servers as these servers consider their time reference directly from the authoritative time source, i.e. UTC (NPLI). Time synchronization plays an important role in the cybersecurity of the nation by enabling the analysis and correlation of events from multiple sources and reconstructing the sequence of events. It ensures reliable authentication, enhances network security, coordinates actions across distributed components, supports the monitoring and control of critical infrastructure, facilitates compliance and auditing and ensures the proper sequencing and validation of transactions To augment and strengthen the cybersecurity in the country, the Indian Computer Emergency Response Team (CERT-In) has issued a directive to all service providers, intermediaries, data centres, body corporates and government organizations to synchronize their ICT infrastructure to the NTP servers of National Physical Laboratory (NPL), National Informatics Centre (NIC), or accurate and standard time sources other than NPL and NIC. Hence, the performance of stratum 1 NTP servers with respect to UTC (NPLI) is important to evaluate and report in public domain. In this paper, the performance of commercially available NTP servers at CSIR -NPL which holds the designation of stratum 1 in the network hierarchy is evaluated with respect to UTC (NPLI). The time offset of the stratum 1 NTP server with respect to the UTC (NPLI) in the laboratory is analysed

D. S. Yadav (✉) · D. Sharma · P. Kandpal · A. Agarwal
CSIR—National Physical Laboratory, New Delhi, India
e-mail: divya.yadav@nplindia.org

D. S. Yadav · D. Sharma · A. Agarwal
Academy of Scientific and Innovative Research (AcSIR), Ghaziabad, India

S. Yadav et al. (eds.), *Advances in Metrology*, Lecture Notes in Mechanical Engineering,
https://doi.org/10.1007/978-981-96-6418-4_5

and reported. The study explores the time synchronization behaviour of stratum 1 NTP server, which may be useful for organizations which are taking time synchronization services from stratum 1 NTP servers which are responsible for nation-wide time synchronization.

Keywords NTP server · Stratum 1 · CERT–In · Cybersecurity · CSIR-NPL · Time synchronization · ICT infrastructure

1 Introduction

The National Physical Laboratory (NPL) maintains the physical realization of Coordinated Universal Time (UTC), which is the primary standard for time and frequency. The physical realization of UTC in India is called UTC (NPLI) [1]. Accurate time synchronization is essential for various applications, including network communication, financial transactions and cybersecurity [2]. NTP servers play a crucial role in maintaining synchronization by providing precise time information to client devices [3–5]. NTP is a network protocol; hence many governments and private agencies such as data centres, service providers, intermediaries, body corporate, financial institutes and IT infrastructure are using NTP servers in their network premises for time synchronization of their device clocks. Assessing the performance of NTP servers is critical to ensure reliable timekeeping across networks [6]. This study introduces a threading-based approach for evaluating NTP server performance on a local network, aiming to enhance time synchronization capabilities and address existing challenges in timekeeping accuracy and establish the concept of network metrology for the measurement of timing device available on network.

The manuscript comprehensively covers the analysis of NTP server performance on a network with respect to reference NTP server directly traceable to UTC (NPLI). In Sect. 2, the concept of Network Time Protocol is elaborated for basic understanding of the concept. Section 3 describes the measurement setup and network architecture used for the current work and analysis. Section 4 discusses the methodology used for formulating the Python-based software for collecting the measurement data on network. In Sect. 5, the measurement results and discussion based on the retrieved results are presented. Finally, the conclusion summarizes the practical findings and the scalability of the network to WAN.

2 Network Time Protocol

The Network Time Protocol (NTP) is a widely used network protocol for synchronizing the clocks of computer systems and network devices over packet-switched, variable-latency data networks [7–9]. It was designed to synchronize time across a network of devices, ensuring that all devices have a consistent and accurate understanding of the current time. It operates in a hierarchical structure with different layers or strata. At the top of the hierarchy are highly accurate time sources, such as atomic clocks or GPS receivers, designated as stratum 0 servers. These servers provide time to stratum 1 servers, which in turn distribute time to stratum 2 servers, and so on. Each subsequent stratum represents a level of hierarchy further from the primary time source as shown in Fig. 1.

NTP operates on a client–server model where the network device available on the network for time synchronization is called an NTP client. These NTP client synchronize their system clocks by exchanging packets containing timestamps. Any network device synchronizes its clocks to NTP servers by exchanging packets containing timestamps. The client sends a request packet to the server, which responds with a packet containing its own timestamp, which is called the Originate Timestamp (T1). The client then records the arrival time of the response packet, which is called the Destination Timestamp (T4), and uses this information, along with the timestamps from the server when the packet is received at server, which is called the Receive Timestamp (T2), and the timestamp when the server sends the response, called the Transmit Timestamp (T3), to calculate the round-trip time and the offset between its clock and the server's clock, as shown in Fig. 2.

The offset of B relative to A is given by

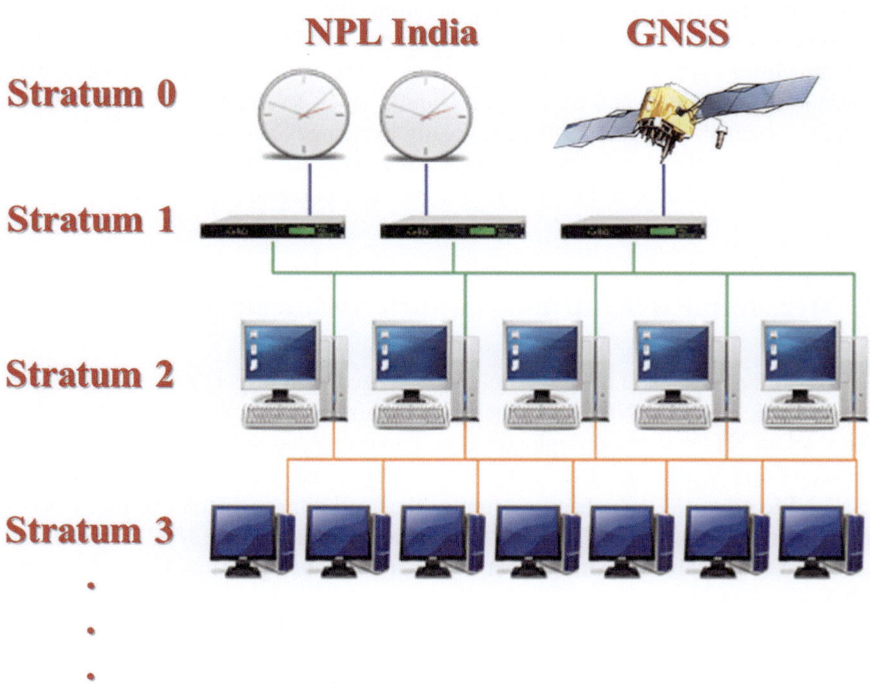

Fig. 1 Hierarchical structure of NTP

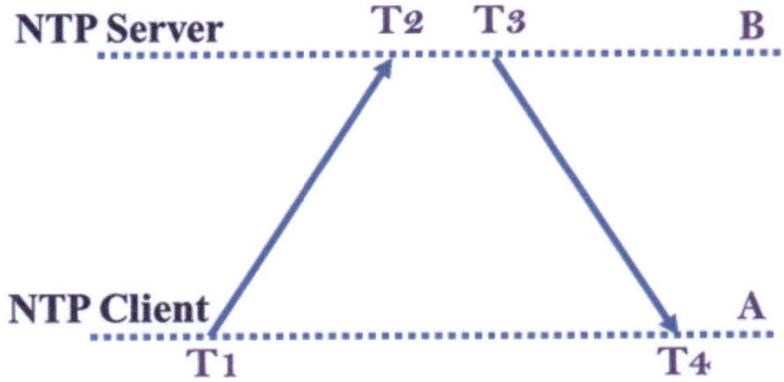

Fig. 2 NTP Packet time stamp exchange

$$\theta = T(B) - T(A) = 1/2 * [(T2 - T1) + (T3 - T4)] \tag{1}$$

and the round-trip delay is given by

$$\Delta = T(ABA) = (T4 - T1) - (T3 - T2) \tag{2}$$

NTP uses sophisticated algorithms to calculate the most accurate time offset and adjust the client's clock accordingly. These algorithms take into account factors such as network latency, jitter and the quality of the time sources to produce a stable and reliable time synchronization. NTP clients select the most appropriate time source based on various criteria, including the server's stratum, reachability and synchronization distance. Clients prefer servers with lower stratum numbers, as they are closer to the primary time sources and typically offer higher accuracy. NTP continuously monitors and adjusts the client's clock to maintain synchronization with the selected time source. It periodically exchanges packets with the server to update the time offset and adjust the clock rate as needed. This continual adjustment ensures that the client's clock remains accurate and consistent over time.

3 Measurement Setup and Network Architecture

CSIR-NPL has a stack of NTP servers synchronized with respect to the authoritative atomic clock, i.e. UTC (NPLI) and attaining stratum 1 status in the network hierarchy. The stack of NTP servers is deployed at two distinct geographical locations, named Primary Time Scale Lab (Lab 211) and Backup Time Scale Lab (Lab 35) in the CSIR-NPL premises, which comes under the Demilitarized Zone in the network architecture. One of the NTP servers is considered the reference NTP server and its performance is evaluated with respect to UTC (NPLI) using a Time Interval Counter (TIC), as shown in Fig. 3.

Since it is a network device, this work demonstrates the evaluation of the stratum 1 NTP servers' performance in terms of time offset and network path delay calculation on the network. Two NTP servers are identified for this analysis from both the labs. The evaluation system is developed using a python-based software tool on a Local

Fig. 3 Measurement setup for reference NTP server

Area Network (LAN), as shown in Fig. 4, which calculates the offset and delay and stores it for further evaluation.

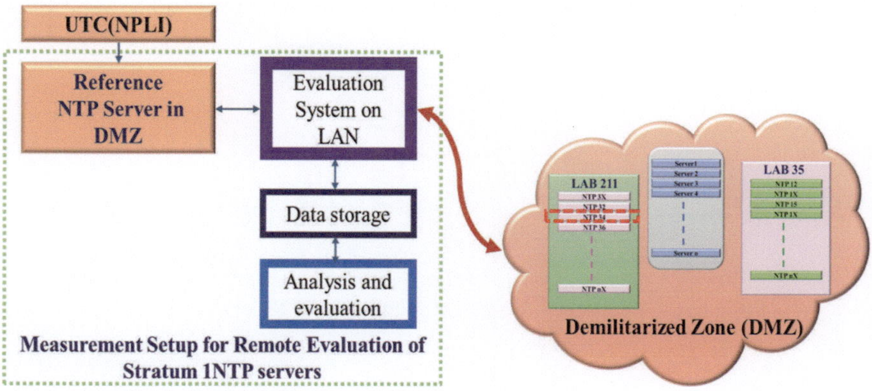

Fig. 4 Measurement setup and network architecture

4 Methodology

To evaluate the performance of a stratum 1 NTP server on a local area network, a method is devised where the computer system acts as the common platform, synchronized with respect to the reference NTP server over the LAN. A software was developed, and the flowchart for this software is shown in Fig. 5.

The computer system sends an NTP request to the reference NTP server and other NTP servers under test for gathering offset and delay data at the same time. The offset represents the difference in time between the respective NTP server clock and the system clock, while the delay is the average time taken for the NTP request to reach to the server and receive the response. The python-based software performs the following tasks simultaneously at every 300 s interval and stores data for further processing.

- **NTP Querying**: The NTP protocol allows the synchronization of time between systems by querying time servers. The program queries multiple NTP servers simultaneously to gather offset and delay data.
- **Threading**: Threading enables concurrent execution of tasks, allowing the program to query multiple NTP servers concurrently. This approach enhances efficiency and reduces the overall time required to collect data from all servers.
- **Data Handling**: Collected data is stored in a CSV file for further analysis.
- **Error Handling**: The program includes robust error handling to manage exceptions gracefully. This ensures that any errors encountered during the querying process are logged and does not halt the execution of the program.

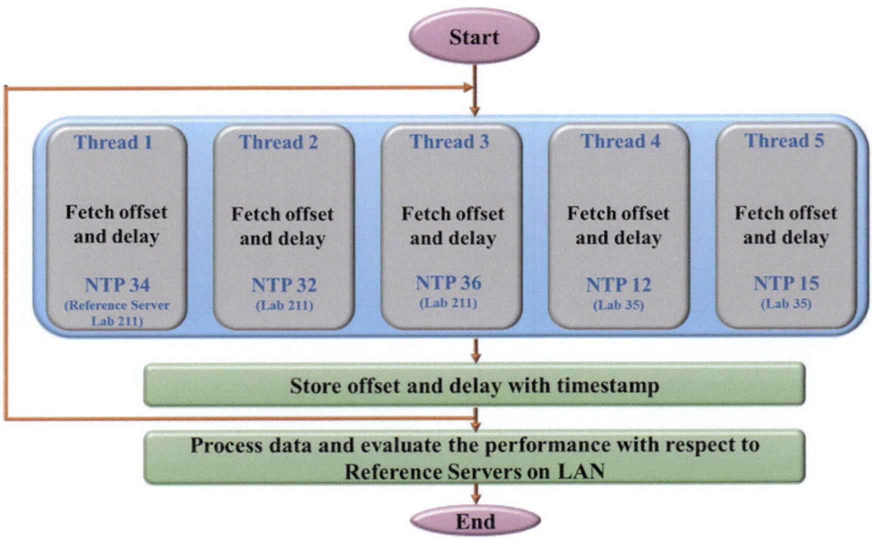

Fig. 5 Methodology flowchart

After gathering the offset and delay data with timestamps, the time offset and network path delay calculation with respect to the reference NTP server are done by eliminating the system clock dependency. The calculation of time offset with respect to the reference NTP server and other Devices under test (DUT) is as follows.

$$Time\ Offset\ at\ timestamp(t)\ with\ respect\ to\ Reference\ NTP\ Server = \theta_{Ref} \quad (3)$$

$$Time\ Offset\ at\ timestamp(t)\ with\ respect\ to\ DUT = \theta_{DUT} \quad (4)$$

To calculate the time offset of an NTP server (DUT) with respect to the reference NTP server, Eq. (3) is subtracted from Eq. (4).

5 Measurement Results and Discussion

In this work, the reference NTP server performance is being evaluated with respect to UTC (NPLI) which is then used to evaluate the performance of other NTP servers on the network.

5.1 Performance of Reference NTP Server w.r.t. UTC (NPLI)

The performance of the reference NTP server is measured with respect to UTC (NPLI) after calculating time offset of 1 PPS output in TIC as shown in Fig. 6. The data are taken every second, and the average Time offset of the reference NTP Server with respect to UTC (NPLI) is **0.7016 μs** and the standard deviation is **0.036 μs**. This measurement provides the traceability link between the Device Under Test (DUT) and UTC (NPLI), as the reference NTP server is evaluated directly with respect to the UTC (NPLI). Furthermore, the reference NTP server is used for evaluating the performance of other NTP servers on LAN.

5.2 Performance of Other NTP Server (DUT) w.r.t. Reference NTP Server

Four NTP servers are used for evaluating the performance on LAN. NTP 32, NTP 36, NTP 12 and NTP 15 are selected from both the labs and their time offset and network path delay are extracted with respect to the reference NTP server using Python-based software, as shown in Fig. 7.

The performance of the stratum 1 NTP server on the network is important for network metrology, where no physical connection exists but the packet analysis is

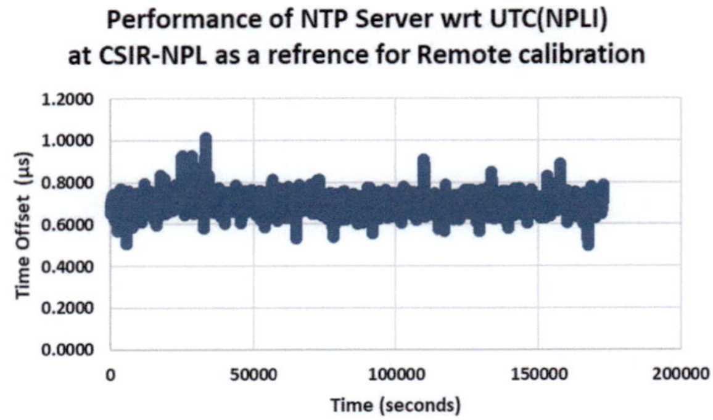

Fig. 6 Performance of reference NTP server with respect to UTC (NPLI) (correct spelling of reference)

required for analysing traceability. Hence, these servers are evaluated on the network and can be traced back to the UTC (NPLI) without the physical measurement of traditional metrology. The statistical parameters are calculated as shown in Table 1.

6 Conclusion and Future Scope

The presented work provides insights into network metrology by utilizing a general-purpose computer system to collect time offset details from a reference NTP server synchronized to UTC (NPLI) within a range of $\pm$ 1 μs. This setup allows for the evaluation of the performance of NTP servers available on a LAN. Despite the limitations in clock precision inherent in general-purpose computer systems, this method effectively establishes traceability to UTC (NPLI). Future studies will involve using a more precise single-board computer, which is expected to enable a more accurate evaluation of NTP server performance.

Fig. 7 Performance of NTP servers (DUT) w.r.t. reference NT server

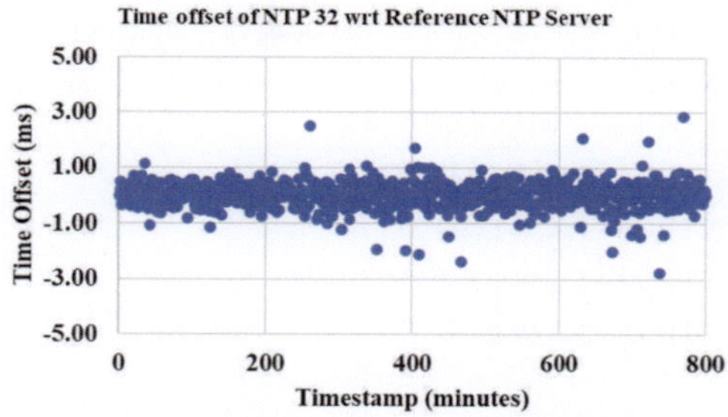

(a) Performance of NTP 32 w.r.t. Reference NTP Server

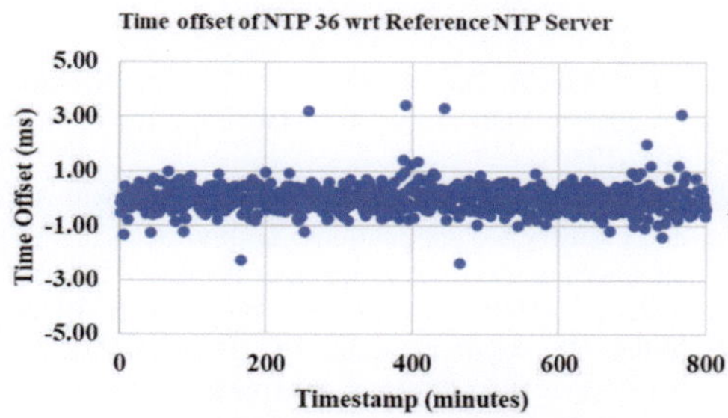

(b) Performance of NTP 36 w.r.t. Reference NTP Server

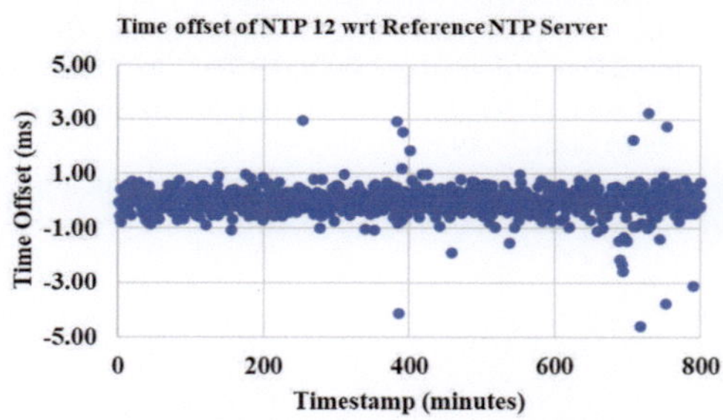

(c) Performance of NTP 12 w.r.t. Reference NTP Server

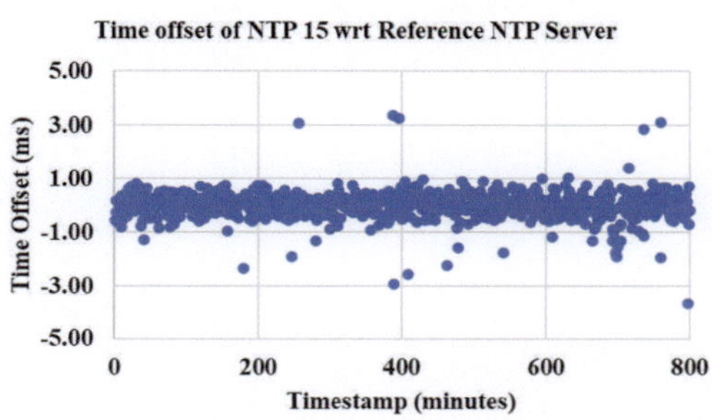

(d) Performance of NTP 15 w.r.t. Reference NTP Server

Table 1 Statistical parameter of device under test

NTP server	Network type and location of the servers deployed	Stratum	Average time offset (ms)	Standard deviation time offset (ms)
NTP 32	LAN–LAB 211	1	0.01	0.48
NTP 36	LAN–LAB 211	1	–0.0129	0.62
NTP 12	LAN–LAB 35	1	–0.02	0.63
NTP 15	LAN–LAB 35	1	0.0033	0.77

References

1. Vattikonda B, Das M, Bhardwaj T, Panja S, Arora P, Gupta A, Aswal DK (2020) "Time and Frequency Metrology." In: Metrology for inclusive growth of India. Springer, Singapore, pp 145–195
2. Sharma D, Yadav DS, Vattikonda B, Agarwal A (2022)Timing accuracy for internet of things and industry 4.0 applications: Technologies and research trends. In: International conference on advances in metrology. Springer Nature Singapore, Singapore, pp 25–34
3. Mills DL (1985) Network time protocol (NTP). No. rfc958
4. Yadav DS, Agarwal A, Jha RC, Dwivedi A (2023) Indian standard time synchronization Via NTP server over different networks. In: Recent advances in metrology. Springer, Singapore, pp 183–192
5. Mounabhargav P, Yadav DS, Sharma D et al (2024) Analysis of ethernet traffic patterns on NTP servers at CSIR NPL. Mapan. https://doi.org/10.1007/s12647-024-00755-0
6. https://www.cert-in.org.in/PDF/CERT-In_Directions_70B_28.04.2022.pdf
7. Mills DL (1991) Internet time synchronization: the network time protocol. IEEE Trans Commun 39(10):1482–1493
8. Mills DL (2006) Computer network time synchronization: the network time protocol. CRC Press
9. Rybaczyk P (2006) Expert network time protocol: An experience in time with NTP. Apress.

Data Acquisition and Remote Monitoring of Critical Parameters of Primary Frequency Standard

Anjali Bisht, Navraj Poudel, Shagufta Rasheed, Manoj Das, Suchi Yadav, Amitava Sen Gupta, and Poonam Arora

Abstract The NPLI-CsF1 is an indigenously developed primary frequency standard for the realization of SI second. The entire system consists of different parts which gives over 30 parameters to be monitored. For accurate and reliable operation of the CsF1 (Acharya et al. in Mapan 32:67–76, 2017), a system is required which logs all the parameters simultaneously and displays it on a single window so that all the parameters can be monitored and one can have a remote access to it. This chapter discusses, as a first step, the design and implementation of a comprehensive 8 channel data logger system (8 parameters) tailored for the NPLI-CsF1 and its integration into the system. In future, we hope to expand the system for all the parameters required for the uninterrupted and precise operation of the primary frequency standard.

Keywords NPLI-CsF1 · Primary frequency standard · SI second · Data logger · Remote monitoring · Real-time data acquisition

1 Introduction

Accurate timekeeping is essential for various applications, including telecommunications, GNSS like GPS, financial markets, and scientific research. It ensures efficient data transfer, precise navigation, reliable financial transactions, and synchronized scientific experiments. Primary frequency standards like the NPLI-CsF1 are crucial for maintaining this high level of precision.

The NPLI-CsF1 clock, developed at CSIR-NPL, is one of the few primary frequency standards globally recognized for realizing the quantum definition of the

A. Bisht · N. Poudel · S. Rasheed · M. Das · S. Yadav · A. S. Gupta · P. Arora (✉)
CSIR-National Physical Laboratory, New Delhi 110012, India
e-mail: arorap.nplindia@csir.res.in

A. Bisht · N. Poudel · M. Das · P. Arora
Academy of Scientific and Innovative Research (AcSIR), Ghaziabad 201002, India

A. S. Gupta
The North Cap University, Gurugram 122017, India

S. Yadav et al. (eds.), *Advances in Metrology*, Lecture Notes in Mechanical Engineering,
https://doi.org/10.1007/978-981-96-6418-4_6

SI second, defined as the unperturbed ground state hyperfine transition frequency of Cs-133 atom, to be 9,192,631,770 when expressed in unit of Hz, which is equal to s^{-1}.The NPLI-CsF1 clock has been designed and developed at CSIR-NPL and contributed to TAI in between 2013 and 2014 and has been approved as primary frequency standard in 2014 [1]. It is a precision primary frequency standard and has been developed by only 10 countries till date. It has been revived, evaluated and is being upgraded for continuous and reliable operation.

Cs Fountain operates in a cyclic and sequential manner [1]. In one cycle, Cs atoms are laser cooled, trapped, launched, interrogated and detected, with each cycle lasting two seconds. This cyclic nature demands a data logger system capable of capturing all relevant parameters such as laser power levels, temperature, humidity, atomic signal sensor readings other critical metrics that influence the operation of the caesium fountain, synchronously. The data logger system's fundamental design serves as an 8-channel power meter for cooling and re-pump lasers. The primary aim is to develop a system that can log these parameters, transmit the data to a remote server, and display the information on a single window for real-time monitoring and analysis. By leveraging components like the ADS1115 ADC for precise analog-to-digital conversion which has I2C (Inter-Integrated Circuit) protocol for communication with master controller and an inbuilt multiplexer, and the ESP8266 Wi-Fi module for efficient data transmission, the system logs data, stores it on a remote server, and displays it via a PHP-based GUI [2–5]. This setup provides precise monitoring and remote access, with future expansions planned for comprehensive parameter coverage.

2 Incentive and Requirement

The continuous operation of the CsF1 primary frequency standard is imperative for its contribution to TAI, demanding a robust and reliable system that ensures uninterrupted functionality over extended periods (at least 10 days). This system must provide real-time data monitoring and remote accessibility for swift troubleshooting to ensure uninterrupted operation and mitigate dead time uncertainty.

Monitoring critical parameters such as laser powers, optical alignment, c-field current sources, laser locking, AOM efficiencies, Pressure gauge current, and environmental conditions (temperature, humidity) is essential to maintain precision. The system's design is intricately synchronized with the Cs Fountain's cyclic operations, ensuring timely data acquisition and analysis. Moreover, the diverse scales and units of these parameters necessitate a tailored, indigenous solution to effectively meet these stringent operational requirements.

One of the primary requirements for seamless operation is to monitor the laser power used for the cooling, launch, and detection beams. These beams are coupled to optical fibers on the optics table and then routed to the physics package, where the cooling, launch, and detection of cesium atoms occur [1]. At this location, the beams are expanded using our indigenously developed beam expanders. A fraction of this power is captured using a photodiode, which outputs a signal in millivolts (mV). This signal is displayed on a power monitor, and the voltage is calibrated to determine the corresponding power in milliwatts (mW).

Monitoring these power levels is crucial for the formation of the magneto-optical trap (MOT) and for obtaining consistent detection return signals essential for clock operation. This setup involves a total of eight laser beams, each requiring individual power monitoring. To manage this task efficiently, we are developing an 8-channel datalogger system.

3 8-channel Data-Logger System

3.1 Hardware Design and Implementation

3.1.1 System Architecture

The basic design of the data logger system is for the purpose of 8-channel power monitor of cooling and re-pump lasers, integrating ADS1115 and ESP8266 to work synchronously with the Cs Fountain cycle. Wi-Fi ESP8266 chip logs the data using Arduino Integrated Development Environment (IDE), a straightforward platform that enables coding in C. ADS1115 handles analog-to-digital conversion and channels data to the ESP8266 synchronously with the TTL signal from the Cs Fountain sequence controller [1]. The ESP8266 commands ADS1115 to convert and send data, which is then stored on a remote PHP server via Wi-Fi. The data is visualized using Dash, a Python library, and displayed through an interactive dashboard. Two ADS1115 are used in parallel for 8 channels (A0, A1, A2, A3 on each chip) [2, 3].

The basic design of the setup is shown in Figs. 1 and 2.

3.1.2 Data Flow

1. Analog Signal Collection: Sensors and photodiodes collect 8 analog signals related to laser cooling, re-pump, and detection laser powers, displayed on an indigenously developed power monitor display unit.
2. Analog-to-Digital Conversion: The ADS1115 converts these analog signals to digital data [2].
3. Data Transmission: The ESP8266 receives the digital data from ADS1115 and transmits it to a remote server via Wi-Fi [3].
4. Data Storage and Display: Upon receiving data from the ESP8266, the PHP application logs the data into a database for long-term storage and analysis. The database structure is designed to efficiently manage and retrieve data. Data is displayed in interactive charts and graphs, enabling users to analyze trends and patterns.

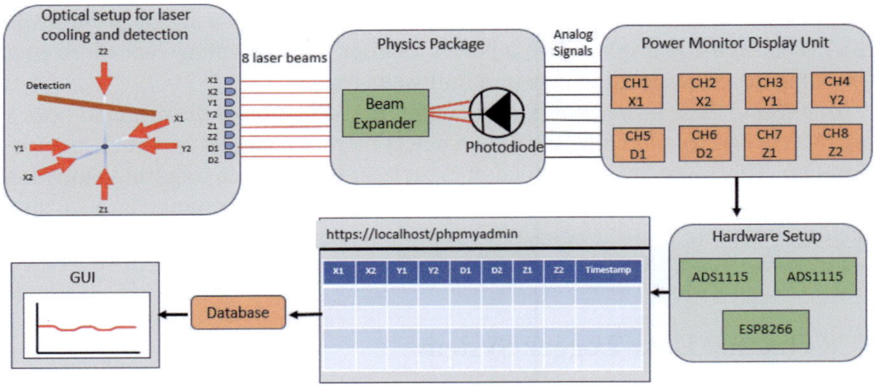

Fig. 1 The flow chart illustrates the optical setup where 8 beams are coupled into optical fibers and sent to the physics package. Each beam is expanded, with a portion of the beam directed to photodiodes. Analog signals from the photodiodes are sent to a power monitor display unit with channels CH1 to CH8. The data is then processed through a hardware setup featuring ADS1115 and ESP8266 components. A database interface, accessible via a URL, displays data columns labeled X1, X2, Y1, Y2, D1, D2, Z1, Z2, and Timestamp. A GUI with a graph is linked to the database.

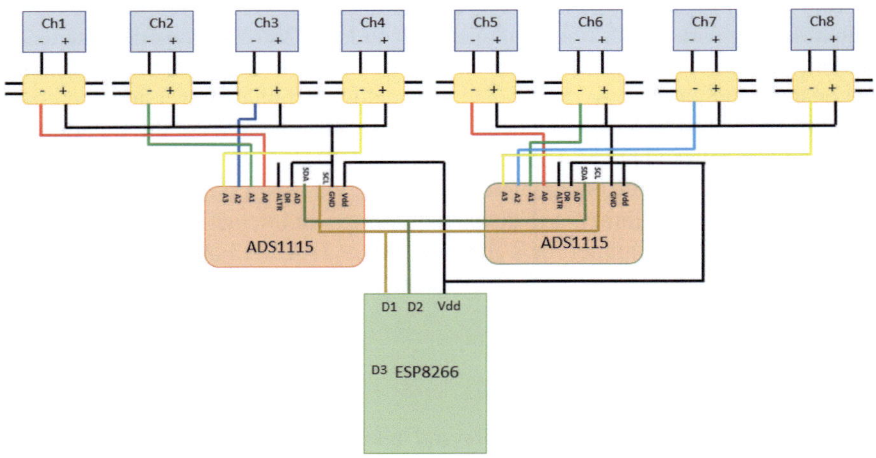

Fig. 2 The figure shows the schematic of the 8-channel data logger system, where each channel (Ch1 to Ch8) has positive and negative terminals, with the positive terminals grounded and negative voltage signals converted to positive values for GUI display through code logic. These channels connect to two ADS1115 analog-to-digital converters, which are linked to an ESP8266 microcontroller via connections labeled D1, D2, D3, and Vdd. Various colored lines represent different connections between the components

5. Synchronization with Fountain Cycle: The ESP8266 is programmed to ensure that data logging occurs every two seconds, synchronized with the Cs Fountain cycle. This synchronization is critical for accurate monitoring and analysis.

3.1.3 Detailed Component Functionality

ADS1115 Analog-To-Digital Converter

Given our need to measure millivolt-level voltages, we chose the ADS1115, a 16-bit analog-to-digital converter (ADC). It provides precise signal monitoring, with features like high-resolution measurement, an integrated multiplexer, and an I2C interface. The ADC's programmable gain amplifier (PGA) supports various inputs across four channels and offers low power consumption. Its internal components, such as a voltage reference, clock oscillator, and digital comparator, simplify integration, and reduce the need for additional external parts [2].

ESP8266 Wi-Fi Module

The ESP8266 is selected for its integration of Wi-Fi capabilities and microcontroller functionality, making it ideal for wireless data transmission applications [3–5]. It effectively interfaces with peripherals like the ADS1115 ADC via its I2C interface, offering a reliable solution for data acquisition and transmission in embedded systems.

4 Software Development

Arduino Code

The Arduino IDE is used to write and upload code to the ESP8266. The code includes.

- Wi-Fi Setup: Configuring the ESP8266 to connect to a Wi-Fi network.
- I2C Communication: Initializing and managing communication with ADS1115.
- Data Handling: Reading data from ADS1115 and preparing it for transmission.
- Synchronization: Ensuring data logging occurs in sync with the fountain cycle (Fig. 3).

This Arduino code is designed for an ESP8266 microcontroller to read analog sensor values using two ADS1115 analog-to-digital converters and transmit the data to a server via HTTP POST requests. The code imports necessary libraries (ESP8266WiFi, ESP8266HTTPClient, WiFiClient, Wire, and Adafruit_ADS1015) and sets up Wi-Fi credentials and server URL. It initializes global arrays for storing sensor readings and defines ADC instances for interfacing with two ADS1115 devices on different I2C addresses. In the setup function, Wi-Fi and serial communication are initialized, ADCs are started, a trigger pin is set as an input, and an interrupt is attached to reset the data index.

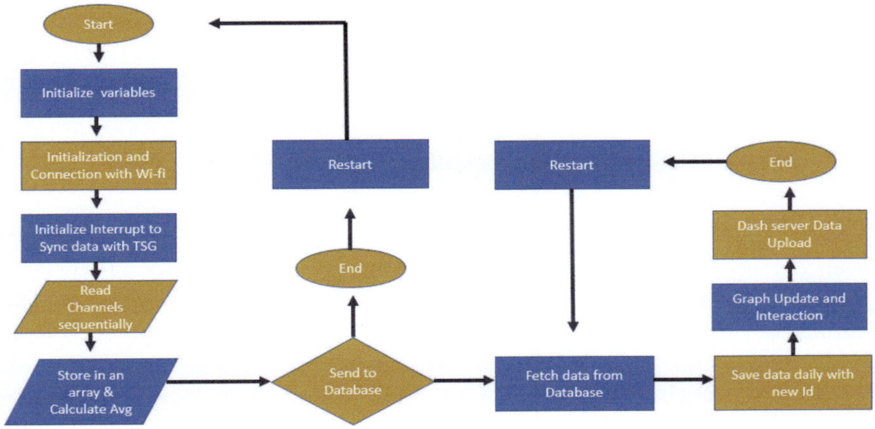

Fig. 3 Flowchart of program algorithm

The loop function continuously reads ADC values, processes them, stores results in arrays, and prints them to the serial monitor. After collecting a certain number of samples, it sends the maximum values from each array to the server using an HTTP POST request. The sampleMax function is used to find the maximum value in each array, ensuring the most significant data is transmitted. The interrupt service routine (ISR) resets the index variable upon a trigger event, allowing the system to start a new data collection cycle. This setup effectively leverages the ESP8266's Wi-Fi and microcontroller capabilities for continuous monitoring and efficient data transmission in embedded applications.

PHP GUI Development

A graphical user interface (GUI) is developed using PHP to display all logged parameters in a single window. The Arduino Integrated Development Environment (IDE) enables the data acquisition and control processes using the ESP8266 and Adafruit ADS1115 libraries. Data collected by the ADS1115 ADCs is transmitted wirelessly via the ESP8266 to a server. Automated Python scripts, fetch and store data in CSV files, reducing manual errors. The Python Dash application then reads, filters, and visualizes this data, providing real-time updates, interactive displays, date range selections, and download options.

- Real-Time Data Display: Shows all monitored parameters in real-time.
- Remote Access: Accessible from any device with internet connectivity.
- User-Friendly Interface: Intuitive design for easy navigation and monitoring.
- Data Analysis Tools: Basic tools for analyzing trends and anomalies in the logged data.

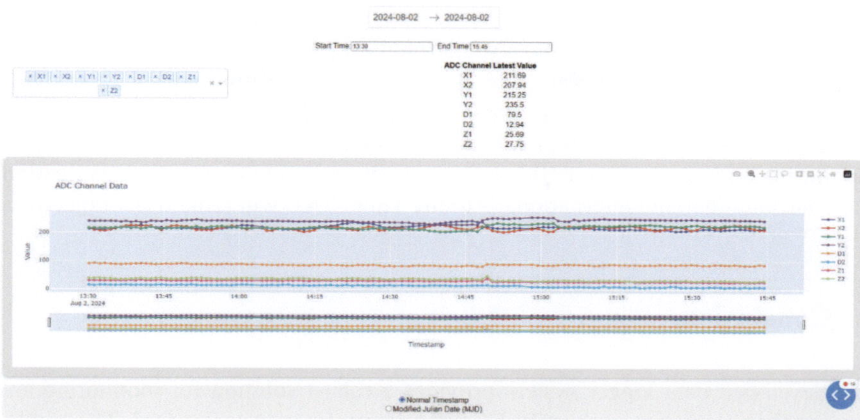

Fig. 4 The figure shows a screenshot of the Dash interface displaying the visual representation of the data logged from the eight channels

This comprehensive setup ensures accurate data acquisition, processing, and analysis through an interactive web interface, facilitating effective monitoring and real time visualization of the 8-channel data.

5 Current Developments

The figure illustrates the data visualization for all eight channels of the datalogger system, which monitors laser powers and is displayed on a PHP based GUI. This data, recorded using Arduino code, is measured in millivolts (mV).

The HOME page features a real-time graph of the eight channels, refreshed every 30 s as shown in Fig. 4. On the Logger page, users can retrieve data by selecting specific start and end dates and times, with options to view data in real time, standard date format, or Modified Julian Date (MJD). This page also displays the most recent values for the eight channels in a table. It also allows users to choose one or more channels for visualization. The interface can be further customized to meet individual needs.

6 Way Ahead/Prospective Advancements

The current setup monitors eight channels, but future work aims to expand input channels by adding more ADS1115 modules to accommodate additional sensors. Enhanced data analysis tools will be incorporated into the GUI, and synchronization algorithms will be refined to improve accuracy. Future enhancements for the project include developing an SMS alert system using a GSM module, integrating email notifications for more comprehensive alerting, creating a mobile app to provide a user-friendly interface for managing alerts, and incorporating advanced analytics and machine learning to predict potential threshold breaches and issue proactive alerts. These improvements will enhance the system's functionality by boosting its monitoring and alerting capabilities. Continuous sustainable operation, evaluation, and contribution to the International Atomic Time (TAI) will remain key objectives.

7 Conclusion

The developed data logger system provides a robust solution for monitoring and logging the parameters of the NPLI-CsF1 primary frequency standard. The combination of the ESP8266 Wi-Fi module and the ADS1115 ADC offers a powerful and efficient solution for applications requiring precise analog-to-digital conversion and wireless data transmission. Python enabled real-time visualization of voltage readings with interactive graphs, improving data analysis and monitoring. The programmability, low power consumption, and Arduino IDE compatibility of the ESP8266, paired with the high-resolution capabilities of the ADS1115, make this setup ideal for continuous monitoring and data logging tasks in embedded systems. Future expansions will enhance the system's capabilities, ensuring comprehensive monitoring and uninterrupted operation of the primary frequency standard.

Acknowledgements This detailed chapter provides an in-depth look at the development, implementation, and testing of a data logger system for the NPLI-CsF1, emphasizing the roles of ADS1115 and ESP8266 in ensuring accurate and reliable operation.

References

1. Acharya A, Bharath V, Arora P, Yadav S, Agarwal A, Gupta AS (2017) Systematic uncertainty evaluation of the cesium fountain primary frequency standard at NPL India. Mapan 32:67–76. https://doi.org/10.1007/s12647-016-0190-4
2. https://www.ti.com/lit/ds/symlink/ads1115.pdf
3. https://www.espressif.com/sites/default/files/documentation/0aesp8266ex_datasheet_en.pdf
4. Abdullah E, Shuhaimi NI, Mohamad R, Kamaluddin NA, Hidayat NM, Burham N (2021) Development of real-time energy monitoring system and data log using NodeMCU ESP 8266 and MYSQL database
5. Schwartz M (2016) Internet of Things with ESP8266. Packt Publishing Ltd

Microplastics in Soil: An Emerging Threat

Anuradha Singh, Prerna Singh, and Manoj Kumar

Abstract Microplastic pollution in soil is an emerging environmental concern with widespread implications for ecosystem health and human well-being. Originating from various sources such as agriculture, industrial activities, and atmospheric fallout, microplastics (MPs) persist in soil environment, altering soil chemistry, physical properties, and biological interactions. These minute plastic particles, predominantly composed of polyethylene (PE) and polystyrene (PS), accumulate in soil as various shapes like films, spheres, beads, and fragments posing risks to soil biota and potentially entering the food chain. Current research highlights the need for standardized methods in sampling, pre-treatment, and analysis to accurately quantify and characterize MPs in soil. Techniques such as extraction of MPs from the soil, including digestion of organic matter, density separation, and chemical characterization with spectroscopic analysis (e.g., Fourier Transform Infrared spectroscopy (FTIR), Raman spectroscopy), are essential in enhancing scientific understanding of MPs pollution in soil. Studies indicate that MPs can disrupt soil nutrient cycling, microbial diversity, and plant growth, which affect agricultural productivity and ecosystem stability. Despite increasing awareness, there are gaps in understanding the long-term impacts of MPs on soil health and biodiversity. Addressing this issue requires collaborative efforts in monitoring, implementing mitigation strategies, and conducting additional research to formulate sustainable solutions. Establishing standardized protocols for assessing MPs in soil is important for comprehensive risk evaluation and efficient environmental management.

Keywords Biodiversity · Environmental impact · Microplastics · Soil health · Soil pollution

---

A. Singh · P. Singh · M. Kumar (✉)
Environmental Sciences & Biomedical Metrology Division, CSIR-National Physical Laboratory, New Delhi, India
e-mail: bhu.manoj@gmail.com; manojkumar.nplindia@csir.res.in

Academy of Scientific and Innovative Research (AcSIR), Uttar Pradesh, Ghaziabad, India

S. Yadav et al. (eds.), *Advances in Metrology*, Lecture Notes in Mechanical Engineering,
https://doi.org/10.1007/978-981-96-6418-4_7

1 Introduction

Plastic is a type of material made from polymers, which are semi-synthetic organic compounds. It is widely used across industries, agriculture, and medicine due to its affordability, strength, and flexibility [1]. Plastic was first invented in 1860, to address the need for durable, lightweight, and versatile materials. But in 2022, the worldwide production of plastic reached a surprising 400.3 million tons [2]. This immense quantity highlights the extensive production, use, and demand for plastic across various sectors, often resulting in inappropriate disposal. This contributes to pollution and presents significant threats to ecosystems. Plastic waste is often dumped, buried or discarded in soil environment, where it persists for long periods, harming the natural balance of ecosystems. These larger plastics degrade due to environmental factors like UV rays, temperature, biological activities, and physical abrasion into tiny plastic particles called microplastics (MPs). The term "microplastics" was coined in 2004 which measure less than 5 mm in size and exhibiting various shapes like fibers, pellets, and foams [1]. MPs pollution in soil ranks as the second most serious scientific concern in environmental and ecological studies. Each year, the quantity of MPs found in cultivated soil surpasses that found in the ocean, indicating soil as a larger reservoir for MPs compared to the ocean [3]. Studies suggested that the concentration of MPs on land could be 4 to 23 times greater than in the ocean [4]. MPs have significant effects on the chemical and physical properties of soil and biota, including microorganisms, plants, and invertebrates [3]. Based on studies polyethylene (PE), polystyrene (PS), and polypropylene (PP) are the three major polymers found in soil. These polymers can disrupt nutrient cycling processes and alter soil pH levels. Because of their minute size, MPs have the potential to be ingested by humans and organisms via food, making them a significant pollutant [5]. There can be various sources of MPs in the soil environment, like agricultural activities, sewage sludge, irrigation water, industrial activities, automobiles, and atmospheric fallout [6].

2 Types of Microplastics (MPs)

MPs are mainly classified into two categories, i.e., primary and secondary. Primary MPs are intentionally manufactured in small sizes for products like makeup, detergents, and plastic fibers (Nylon). Once released into the environment, MPs persist for extended periods due to their low degradation rate [7]. Secondary MPs, on the other hand, form when bigger plastic breaks down into tiny pieces by wind abrasion, biodegradation, and ultraviolet radiation from sunlight [7]. MPs find their way into soil through various routes such as agricultural films, irrigation water, compost, sewage sludge, atmospheric deposition, traffic, and waste disposal methods (Fig. 1). Agricultural films are made up of polyvinyl chloride and polyethylene and are used to regulate the soil temperature for the growth of plants [8]. Due to their non-biodegradable nature, it is an important source of MPs pollution in agricultural soil. Furthermore, sludge, along with compost is a significant source of MPs. Both are mostly used to enhance soil fertility, improving the chemical and physical properties of the soil by providing essential nutrients such as nitrogen and phosphorus [9–11]. Current treatment and technologies of sludge and composting fail to remove MPs, making them major sources of MPs in soil. Irrigation water for farming can come from treated sewage water, surface water like rivers and lakes, or groundwater from underground. Around 20 million hectares of farmland use sewage water that has not been treated [12]. Cosmetics like makeup and washing powder are causing MPs, to end up in household wastewater. The main material in these MPs is from products like face wash and body wash is polyethylene [13]. Many studies have revealed that a significant amount of MPs deposited in soil is through surface runoff. According to the study conducted by Kim et al. [14], soil environment near to roads could be contaminated by vehicle tire wear. Moreover, vehicle tire wear globally emits an estimated 0.81 kg of MPs per person each year, predominantly composed of synthetic rubber [15, 16]. The primary pathways through which MPs enter the soil environment include atmospheric fallout and run-off water. Furthermore, atmospheric deposition is the another source of MPs contamination in soil. Due to their lower density, MPs easily settle into the soil, making it a larger sink for these particles compared to the oceans [6]. Studies have also identified various types of MPs in the atmospheric environment, existing in multiple forms, which may also serve as a significant source of MPs contamination in soil through deposition [17].

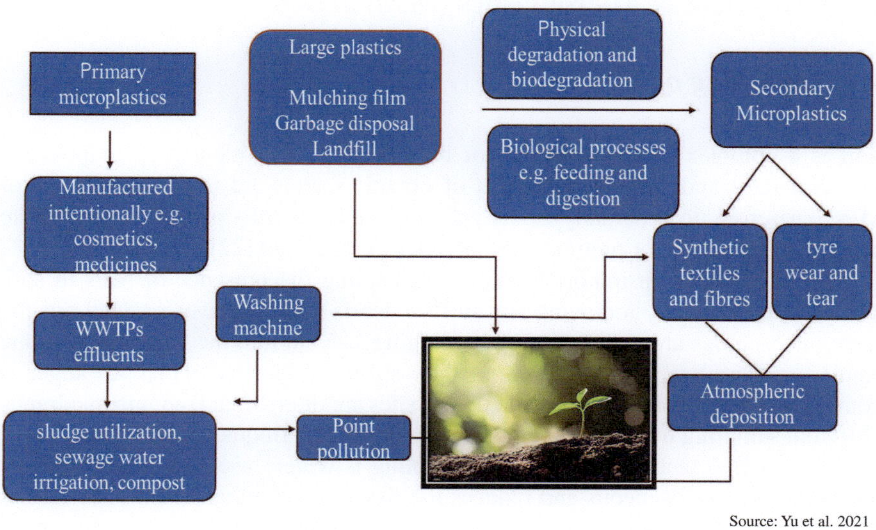

Fig. 1 Sources of microplastics in soil (*Source* Extracted by Yu et al. [1])

3 Transportation of Microplastics (MPs) in Soil

After detecting the presence of MPs in soil, it is crucial to study their transportation across the soil layer. MPs in soil can move around based on their size, composition, shape, and aging. Smaller particles with more polymer composition move faster than larger ones [18]. Fibrous MPs which are predominantly observed in soil after atmospheric fallout, easily travel from soil to shoreline areas [19]. Aged plastic tends to move in vertical direction. The complexity of soil plays a significant role in determining how MPs move within it. Because soil is porous in nature having many tiny pores that allow liquids and small particles to pass through. MPs migrate through these pores and potentially contaminate groundwater [20]. Soil provides an essential habitat for a variety of species, including microscopic microbes and earthworms. This biodiversity is crucial for soil health and ecosystem function. Earthworms, also known as "farmer's friends," improve soil through mixing, aeration, and decomposition. However, they can also swallow and carry these MPs particles deeper into the soil and food web, thereby threatening human health [21]. This establishes a miniature plastic loop within ecosystems. Carlin et al. [5] stated that predators at the top of the food chain carry more MPs than species present at the bottom. As these plastics move up the food chain from lower trophic level to the higher trophic level, they can accumulate in higher concentrations in animals at the top, posing potential risks to their health and possibly impacting humans who consume them.

4 Assessment of Microplastics (MPs) in Soil

4.1 Sampling of MPs

Soil is a complex matrix made up of both, organic and inorganic components. The structural and surface properties of organic matter aid in the adsorption of MPs. Therefore, to accurately detect and analyze these MPs particles, appropriate sampling, extraction, or pretreatment methods are required [22].

Soil sampling is an important step in estimating the presence of MPs in soil. The process begins by removing any surface debris such as leaves or dead organic material. Samples are then collected by digging 1–5 cm into the soil, though some studies may go deeper, depending on their specific objectives [11]. An auger is commonly used for sampling, and sampling sites are often organized into quadrants. Different sampling methods are employed, with studies frequently using random or zigzag sampling techniques. Random sampling is widely used, as it helps ensure the sample is homogeneous and reduces the risk of contamination [15]. To further minimize contamination, it is recommended to collect soil samples in paper bags or glass jars. These samples are often dried to remove moisture which can be done either by oven-drying at 60 -70 C or by air-drying at room temperature for 2–3 days

[22]. Both methods are valid, though air-drying is sometimes preferred for maintaining sample integrity. After sampling, the soil samples undergo pretreatment, which involves processes like sieving, density separation, and chemical digestion to isolate microplastics from other soil components [1].

4.2 Extraction of Microplastics (MPs) from the Soil

The pretreatment process begins with sieving the dried soil to remove particles larger than 5 mm in size [22]. For quantifying MPs, it is important to extract them from the soil aggregates through a process known as pre digestion. Various methods are used for pre digestion, including acids, Fenton's reagent, alkalis, enzymes. These substances break down and digest organic matter, releasing MPs. However, due to the delicate nature of MPs, acids are generally avoided as they can distort their shape [23, 24]. According to Li et al. [25], hydrogen peroxide (H_2O_2) is effective in oxidizing organic matter and releasing MPs from the aggregates. Hurley et al. [26] conducted numerous experiments and reported that Fenton's reagent is particularly effective for extracting MPs from various samples. Currently, there is no universally accepted standard method for extracting MPs from soil, and researchers continue to explore different approaches designed to various soil types and conditions [6].

After digesting the soil samples, the next step is to separate the components by density. This process is based on the principle that polymers have lower density than a saturated salt solution, which enables effective separation. Various salts such as NaCl, $CaCl_2$, $ZnCl_2$, NaBr, and NaI are used for this purpose. A saturated solution of these salts is prepared and mixed with the extracted or digested samples. MPs, which typically have lower densities than salt solution, float on the supernatant and are filtered out using vacuum filtration [27]. MPs are present in various types and density ranges; therefore, depending on the specific type of MPs or polymer intended for extraction, saturated salt solution is prepared. This solution must have a density greater than that of the polymer being extracted to ensure effective separation. Among them, NaCl is preferred due to its effectiveness, cheap and safety compared to alternatives like $ZnCl_2$ (which is toxic), NaBr (which is expensive), and NaI (which is also toxic). This method allows for the separation and recovery of MPs from soil samples based on their density differences relative to soil particles [28]. In cases where aggregates are not easily broken down, several other separation techniques can be used to extract MPs. These techniques are electrostatic separation which uses electric charge to separate particles, oil-based separation uses oil to float MPs which can then be filtered out, ultrasonication uses high frequency waves to break aggregates, centrifugation rotates the sample with high speed to separate the samples based on their density and magnetic separation techniques uses magnet to separate the MPs. In some studies, density separation is performed before the digestion of organic matter [1, 28], whereas in others, it is performed afterward. This variability indicates the lack of a standardized method for quantifying and assessing MPs.

4.3 *Identification and Characterization of Microplastics (MPs)*

After density separation, the next step involves staining the filter paper containing the supernatant with Nile Red solution and stereomicroscope is used for the visual identification [29–31]. Nile Red solution is used to stain MPs because of its ability to bind with hydrophobic materials such as plastics. When exposed to blue or ultraviolet light, Nile Red emits a strong red fluorescence. This fluorescence makes it easy to distinguish MPs under a stereomicroscope. This staining helps in the physical or visual characterization of MPs under a stereomicroscope.

Under the stereomicroscope, the shape, size, and color of the MPs are identified and analyzed. This process is important for accurately classifying and understanding the characteristics of MPs particles present in the sample. Physical characterization alone is insufficient for thorough MPs analysis. It can be time-consuming and may misidentify certain MPs without revealing polymer composition [32]. Therefore, chemical analysis techniques are employed to enhance identification and characterization because it accurately identifies MPs, determines polymer composition. Chemical characterization techniques include spectroscopic methods such as FTIR, Raman spectroscopy (Raman), and NIR (Near Infrared Spectroscopy). Among them, FTIR and Raman are extensively utilized for the chemical characterization of MPs. Additionally, various thermal analysis techniques are employed, such as thermogravimetric analysis mass spectrometry (TGA-MS), pyrolysis–gas chromatography mass spectrometry (Pyr-GC–MS), and thermal extraction desorption gas chromatography mass spectrometry (TED-GC–MS) [33–35]. However, each technique has its own merits and limitations. FTIR and Raman spectroscopy are favored among spectroscopic methods due to their widespread applicability. FTIR with ATR (Attenuated Total Reflection) mode is indeed widely used for the identification of MPs, particularly because ATR enhances the ability to analyze samples directly without extensive preparation [36]. Finely digesting organic matter makes it easier for FTIR and Raman spectroscopy to accurately detect and analyze MPs. Raman spectroscopy, though, can be susceptible to fluorescence interference, which limits its utility in some contexts.

5 Worldwide Studies on Microplastics (MPs) in Soil

Microplastics in soil have been extensively studied worldwide, revealed that fragments and fibers are the predominant shapes observed across various regions. The presence of MPs in soil varies significantly depending on land use practices. Furthermore, Liu et al. [37] examined various land use types, including grasslands, drylands, paddy fields, and plastic greenhouses. Their research indicated that drylands tend to have higher concentrations of MPs compared to greenhouse areas. They stated that specific environmental conditions in dryland ecosystems might be the reason of greater accumulation of MPs in the soil. Similarly, Zhang et al. [38] observed that farmlands generally had more MPs than plantations, attributed to different agricultural practices and soil management techniques affecting MPs deposition and retention. Studies on polymer composition consistently identified polypropylene (PP) and polyethylene (PE) as major types of MPs in soil samples, with PE particularly sourced from plastic mulching In China, fibrous MPs were predominantly found in farmland, with lower concentrations observed in orchards and secondary forests, and the least amount detected in plantations [39]. Moreover, MPs abundance in farmland correlated positively with population density, suggesting anthropogenic factors play a significant role in MPs distribution. Overall, these findings highlighted the influence of land use practices on MPs presence in soil. Global studies showed that land use practices, such as in drylands and farmlands, heavily influenced MPs concentrations in soil, dominated by polypropylene and polyethylene from agricultural activities like plastic mulching. Urgent action is needed to implement targeted strategies to mitigate MPs pollution and safeguard soil health and ecosystem balance on a global scale.

6 Studies of Microplastics (MPs) in Soil of India

There are limited studies on MPs pollution in soil reported from India (Table 1). These studies identified predominant polymers such as polyethylene and polystyrene among the MPs particles analyzed. The MPs were commonly observed in the form of fragments and films, indicating diverse sources and degradation patterns. The presence of these polymers in soil suggests a widespread contamination issue affecting various types of land across these regions. This contamination is concerning due to the potential environmental and ecological impacts, including risks to soil health, plant growth and potential ingestion by soil-dwelling organisms. More studies should be there to understand the distribution and characteristics of MPs in soil, for developing effective strategies to mitigate their environmental consequences and safe agricultural and ecological systems [40–44].

7 Impact of Microplastics (MPs) on Physical and Chemical Properties of Soil

MPs, due to their small size and large surface area, adversely affect soil chemical and physical properties by adsorbing harmful particles such as persistent organic pollutants (POPs) and heavy metals. Li et al. [45] conducted a study that examined the impact of MPs on various soil parameters. They observed significant effects on dissolved organic carbon (DOC), dissolved organic nitrogen (DON), dissolved phosphorous, and fulvic acid levels. Additionally, changes were noted in organic phosphorus, phosphates, as well as hydrolase and phenol oxidase activities. De Souza et al.

Table 1 Previous studies of microplastics (MPs) in soil across India

Study area	Land-use type	Counts/ Number of MPs	Polymers of MPs identified	Shapes of MPs identified	References
Indore (Madhya Pradesh)	Agricultural and recreational areas	11 particles/kg and 68.25 particles/kg	Polyethylene terephthalate, polypropylene, polyethylene, polyester, and polyamide	Fibers	[40]
Una (Himachal Pradesh)	Urban, rural, agricultural	NA	NA	Fragments, film, pellet and fibers	[41]
Coimbatore (Tamil Nadu)	Different agricultural sites	NA	Polyethylene, Polyvinyl, alcohol, Polystyrene, polyacrylonitrile, nylon, polypropylene, polyvinyl methyl, and polyvinyl chloride	Fragments and films	[42]
Surat (Gujrat)	Industrial soil, residential soil, road side soil	103.2 ± 57.28 particles 100 g − 1	Polyethylene, polypropylene, and polyethylene, terephthalate	Fragments and fibers, sphere, films	[43]
Bhopal (Madhya Pradesh)	Agricultural land	307.5 ± 9.19 and 69.5 ± 4.95	Polyethylene and propylene	Fibers and fragments	[44]

[46] investigated the effects of commonly found polymers in soil. The study highlighted differences in how each polymer affects soil aggregates: polyester reduced water-stable aggregates, whereas polyethylene increased them. These changes in water-soluble aggregates affect the soil micro-environment. Despite these findings, research on the impact of MPs on soil properties is still in its early stages, making it challenging to draw conclusions about overall effects of MPs on soil.

8 Effect of Microplastics (MPs) on Various Living Organisms

MPs present in the soil pose significant risks to a variety of species. For example, MPs present in soil enter animal's bodies through ingestion, where they can remain in the gut for extended periods, impacting gut health and feeding habits [47]. The digestive systems of animals break down plastic particles into smaller forms that can be absorbed. This ingestion affects reproductive health and disrupts normal biological processes. Such as, face masks emit MPs that affect the spermatogenesis cycle in male earthworms [48]. Additionally, plastics contain harmful substances like toxic heavy metals, antibiotics, and pesticides. These substances leach into animal's bodies when they consume plastic materials, posing further risks to their health and ecosystem stability. Furthermore, they also affect soil microorganisms and biomass activity, as well as nutrient cycling, which in turn impacts seed germination and growth. Over time, these MPs can break down into nanoparticles, which are absorbed by plants and subsequently enter the human body through the food chain. According to Qi et al. [49], the growth of wheat was affected negatively by the presence of LDPE and plastic mulch. Bandmann et al. [50] also found that microbeads in soil enter plant cells through cellular uptake indicating their presence and absorption within plants. In addition, MPs hinder nutrient cycling by clogging pores in the soil. MPs impact soil properties and microorganism diversity which play a major role in nutrient exchange between roots and plants by releasing nutrients, degradation. According to a study by De Souza [46], soils containing microfibers exhibit lower microbial diversity compared to soils without microfibers. Some microorganisms help in plastic degradation but can also absorb harmful chemicals like heavy metals, POPs, and antibiotics from MPs. Moreover, MPs in soil clog pores, limiting oxygen supply and affecting aerobic bacteria. They can also alter the habitat within soil aggregates, further impacting microbial communities. Research on MPs effects on soil microorganisms are limited highlighting the need for further studies to completely understand their ecological implications.

9 Conclusion

In conclusion, MPs pollution in soil presents a significant environmental challenge with far-reaching implications. Originating from various sources including agriculture, industry, and atmospheric deposition, MPs persist in soil ecosystems, affecting soil chemistry, physical properties, and biological communities. Their small size facilitates ingestion by organisms, potentially entering the food chain and impacting human health. Despite ongoing research into extraction and characterization methods, comprehensive understanding of their long-term effects on soil remains limited. Addressing this issue requires concerted efforts in monitoring, mitigation strategies, and further research to safeguard soil health and ecosystem integrity. In conclusion, the widespread presence of MPs in soil presents a significant environmental challenge. Studies consistently highlight polyethylene and polystyrene as the predominant types of MPs found, typically in the form of fragments and films. This contamination poses risks to soil health, agricultural productivity, and ecosystems, emphasizing the urgent need for thorough research and effective mitigation strategies. Despite increasing global awareness and research efforts, the lack of standardized methods for quantifying and characterizing soil MPs complicates accurate assessment of contamination and their ecological impacts. Variations in sampling, pretreatment, and analytical techniques further hinder comprehensive understanding. Moving forward, establishing standardized protocols is important to reliably quantify and characterize MPs in soil. This will enable better risk assessments, deepen our knowledge of MPs behavior in soil, and guide targeted mitigation efforts. Continued research into the ecological and health impacts of MPs on soil organisms is essential for developing sustainable solutions to address this pressing environmental issue.

Acknowledgements The authors are grateful to the Director of CSIR-National Physical Laboratory for providing the infrastructure and support for this study. Anuradha Singh (JRF) and Prerna Singh (JRF) would like to acknowledge the University Grant Commission (UGC), Govt. of India for providing research grants and acknowledges the support of the Academy of Scientific and Innovative Research (AcSIR) in facilitating their Ph.D. studies.

References

1. Yu J, Adingo S, Liu X, Li X, Sun J, Zhang X (2022) Micro plastics in soil ecosystem-A review of sources, fate, and ecological impact. PSE 68(1):1–17
2. Pilapitiya PNT, Ratnayake AS (2024) The world of plastic waste: a review. Clean Mater 100220
3. Horton AA, Walton A, Spurgeon DJ, Lahive E, Svendsen C (2017) Microplastics in freshwater and terrestrial environments: Evaluating the current understanding to identify the knowledge gaps and future research priorities. Sci Total Environ 586:127–141
4. Nizzetto L, Futter M, Langaas S (2020) Are agricultural soils dumps for microplastics of urban origin? Environ Sci Technol 50(20):10777–10779
5. Carlin J, Craig C, Little S, Donnelly M, Fox D, Zhai L, Walters L (2020) Microplastic accumulation in the gastrointestinal tracts in birds of prey in central Florida, USA. Environ Pollut 264:114633
6. Yang H, Yumeng Y, Yu Y, Yinglin H, Fu B, Wang J (2022) Distribution, sources, migration, influence and analytical methods of microplastics in soil ecosystems. Ecotoxicol Environ Saf 243:114009
7. Microplastics (2023) www.hannover-re.com (Accessed on 4–06–24)
8. Singh A, Kaushal A, Garg S, Chawla N (2017) Effect of different coloured plastic mulches on growth, yield and quality of drip fertigated bell pepper (Capsicum annum var. grossum). Indian J Hortic 74(2):292–294
9. Naeini SM, Cook HF (2000) Influence of municipal waste compost amendment on soil water and evaporation. Commun Soil Sci Plant Anal 31(19–20):3147–3161
10. Slater RA, Frederickson J (2001) Composting municipal waste in the UK: some lessons from Europe. RCS 32(3–4):359–374
11. Cherif H, Ayari F, Ouzari H, Marzorati M, Brusetti L, Jedidi N, Hassen A, Daffonchio D (2009) Effects of municipal solid waste compost, farmyard manure and chemical fertilizers on wheat growth, soil composition and soil bacterial characteristics under Tunisian arid climate. Eur J Soil Biol 45:138–145
12. Mintenig SM, Int-Veen I, Loder MG, Primpke S, Gerdts G (2017) Identification of microplastic in effluents of waste water treatment plants using focal plane array-based micro-Fourier-transform infrared imaging. Water Res 108:365–372
13. Lei K, Qiao F, Liu Q, Wei Z, Qi H, Cui S, An L (2017) Microplastics releasing from personal care and cosmetic products in China. Mar Pollut Bull 123(1–2):122–126

14. Kim LH, Kang J, Kayhanian M, Gil KI, Stenstrom MK, Zoh KD (2006) Characteristics of litter waste in highway storm runoff. Water Sci Technol 53:225–323
15. Choi YR, Kim YN, Yoon JH, Dickinson N, Kim KH (2021) Plastic contamination of forest, urban, and agricultural soils: a case study of Yeoju City in the Republic of Korea. JSS 21:1962–1973
16. Kole PJ, Lohr AJ, Van Belleghem FG, Ragas AM (2017) Wear and tear of tyres: a stealthy source of microplastics in the environment. Int J Environ Res Public Health 14(10):1265
17. Dris R, Gasperi J, Rocher V, Saad M, Renault N, Tassin B (2015) Microplastic contamination in an urban area: a case study in Greater Paris. J Environ Chem 12(5):592–599
18. O'Connor D, Pan S, Shen Z, Song Y, Jin Y, Wu WM, Hou D (2019) Microplastics undergo accelerated vertical migration in sand soil due to small size and wet-dry cycles. Environ Pollut 249:527–534
19. Koutnik VS, Leonard J, Alkidim S, DePrima FJ, Ravi S, Hoek EM, Mohanty SK (2021) Distribution of microplastics in soil and freshwater environments: Global analysis and framework for transport modeling. Environ Pollut 274:116552
20. Gao J, Pan S, Li P, Wang L, Hou R, Wu WM, Hou D (2021) Vertical migration of microplastics in porous media: Multiple controlling factors under wet-dry cycling. J Hazard Mater 419:126413
21. Rodriguez-Seijo A, Lourenço J, Rocha-Santos TAP, Da Costa J, Duarte AC, Vala H, Pereira R (2017) Molecular effects of microplastics in Eisenia andrei Bouché. Environ Pollut 220:495–503
22. Radford F, Zapata-Restrepo LM, Horton AA, Hudson MD, Shaw PJ, Williams ID (2021) Developing a systematic method for extraction of microplastics in soils. ANAL METHODS 13(14):1695–1705
23. Catarino AI (2017) Development and optimization of a standard method for extraction of microplastics in mussels by enzyme digestion of soft tissues. Environ Toxicol Chem 36:947–951
24. Claessens M, Van Cauwenberge L, Vandegehuchte MB, Janssen CR (2013) New techniques for the detection of microplastics in sediments and field collected organisms. Mar Pollut Bull 70(1–2):227–233
25. Li Q, Wu J, Zhao X, Gu X, Ji R (2019) Separation and identification of microplastics from soil and sewage sludge. Environ Pollut 254:113076
26. Hurley RR, Lusher AL, Olsen M, Nizzetto L (2018) Validation of a method for extracting microplastics from complex, organic-rich, environmental matrices. Environ Sci Technol 52(13):7409–7417
27. Veerasingam S, Ranjani M, Venkatachalapathy R, Bagaev A, Mukhanov V, Litvinyuk D, Vethamony P (2020) Microplastics in different environmental compartments in India: Analytical methods, distribution, associated contaminants and research needs. TrAC 133:116071
28. Withana PA, Senadheera SS, He D, Bank MS, Gu C, Hwang SY, Ok YS (2024) Soil microplastic analysis: a harmonized methodology. Crit Rev Environ Sci Technol 54(15):1138–1163
29. Corradini F, Meza P, Eguiluz R, Casado F, Huerta-Lwanga E, Geissen V (2019) Evidence of microplastic accumulation in agricultural soils from sewage sludge disposal. Sci Total Environ 671:411–420
30. Scheurer M, Bigalke M (2018) Microplastics in Swiss floodplain soils. Environ Sci Technol 52(6):3591–3598
31. Zhou B, Wang J, Zhang H, Shi H, Fei Y, Huang S, Barceló D (2020) Microplastics in agricultural soils on the coastal plain of Hangzhou Bay, east China: Multiple sources other than plastic mulching film. J Hazard Mater 388:121814
32. Hidalgo-Ruz V, Gutow L, Thompson RC, Thiel M (2012) Microplastics in the marine environment: a review of the methods used for identification and quantification. Environ Sci Technol 46(6):3060–3075
33. David J, Steintmetz Z, Kucerik J, Schaumann GE (2018) Quantitative analysis of poly (ethylene terephthalate) microplastics in soil via thermogravimetry–mass spectrometry. Anal chem 90(15):8793–8799

34. Dumichen E, Eisentraut P, Bannick CG, Barathel AK, Senz R, Braun U (2017) Fast identification of microplastics in complex environmental samples by a thermal degradationmethod. Chemosphere 174:572–584
35. Kappler A, Fischer M, Scholz-Bottcher BM, Oberbeckmann S, Labrenz M, Fischer D, Voit B (2018) Comparison of μ-ATR-FTIR spectroscopy and py-GCMS as identification tools for microplastic particles and fibers isolated from river sediments. ABC 410:5313–5327
36. Kim SK, Kim JS, Lee H, Lee HJ (2021) Abundance and characteristics of microplastics in soils with different agricultural practices: Importance of sources with internal origin and environmental fate. J Hazard Mater 403:12399
37. Liu TN, Yang W, Chang J (2022) Microplastics pollution in the soils of various land-use types along Sheshui River basin of Central China. Sci Total Environ 806:150620
38. Zhang Y, Wang K, Chen W, Ba Y, Khan K, Chen W, Xu L (2022) Effects of land use and landscape on the occurrence and distribution of microplastics in soil. China. Sci Total Environ 847:157598
39. Xu G, Yang L, Xu L, Yang J (2022) Soil microplastic pollution under different land uses in tropics, southwestern China. Chemosphere 289:133176
40. Singh S, Chakma S, Alawa B, Kalyanasundaram M, Diwan V (2024) Microplastic pollution in terrestrial environment: Identification, characterization, and risk assessment in Indore. Central India. Soil Use Manage 40(2):13053
41. Dhiman S, Sharma D, Kotia N, Sinha R (2024) Estimation of microplastics distribution in soil sample from District Una. Himachal Pradesh. J Toxicol Stud 2(1):527–527
42. Karthika S, Sivasubramanian K, Dhevagi P, Thavamani P, Rajkishore SK, Amrutha VM, Lakshmanan A (2024) Distribution pattern and risk assessment of microplastics contamination in different agricultural systems. J Environ Anal Chem 1:20
43. Bhavsar PS, Chovatiya BV, Kamble SB, Gore AH (2024) Extraction and Analysis of Microplastics in the Soil of Diamond City, Surat (Gujarat, India): Ecological Risk, Pollution Indices, and Greenness Evaluation. ACS Agric Sci Technol 4(5):614–625
44. Singh S, Chakma S, Alawa B, Kalyanasundaram M, Diwan V (2024) Assessment of microplastic pollution in agricultural soil of Bhopal, Central India. Journal of Material Cycles and Waste Management, 26(2): J Mater Cycles Waste Manag 708–722
45. Li M, Yu H, Wang Y, Li J, Ma G, Wei X (2020) QSPR models for predicting the adsorption capacity for microplastics of polyethylene, polypropylene and polystyrene. Sci Rep 10(1):14597
46. De Souza Machado AA, Kloas W, Zarfl C, Hempel S, Rillig MC (2018) Microplastics as an emerging threat to terrestrial ecosystems. Glob Change Biol 24(4):1405–1416
47. Boots B, Russell C, Green DS (2019) Effects of microplastics in soil ecosystems: above and below ground. Environ Sci Technol 19:11496–11506
48. Kwak JI, An YJ (2021) Post COVID-19 pandemic: Biofragmentation and soil ecotoxicological effects of microplastics derived from face masks. J Hazard Mater 416:126169
49. Qi Y, Yang X, Pelaez AM, Lwanga EH, Beriot N, Gertsen H, Geissen V (2018) Macro-and micro-plastics in soil-plant system: Effects of plastic mulch film residues on wheat (Triticum aestivum) growth. Sci Total Environ 645:1048–1056
50. Bandmann V, Müller JD, Köhler T, Homann U (2012) Uptake of fluorescent nano beads into BY2-cells involves clathrin-dependent and clathrin-independent endocytosis. FEBS Lett 586(20):3626–3632

Experimental Study of the Metrological Characteristics of an Industrial Hydraulic High-Pressure Transducer Over a Period of Four Years

Raman Kumar Sharma, Afaqul Zafer, Lalit Kumar, Jasveer Singh, Sanjay Yadav, and Nita Dilawar Sharma

Abstract Accurate measurement plays a key role in our daily life. If we can measure correctly, we can save time and energy. For good quality work, accurate and precise pressure measurement is very important in all industries, including the defense and service sectors. Many types of pressure gauges are available at present; however, pressure transducers are one of the most used devices that give direct pressure readings in response to some mechanical deformation. At times, pressure transducers have sensitivity issues, non-linearity, high hysteresis loss and zero error which limits their use in highly precise industrial applications. However, with technological advances and recent developments in pressure metrology, pressure transducers in high-pressure ranges are increasingly being used in the defense and automobile sectors also. In the present study, we have investigated the behavior of an industrial high-pressure transducer over a period of four years. Our experiments have compared a strain gauge-type transducer against the national hydraulic pressure standard. This transducer converts pressure into mechanical strain on the strain-generating body, and the mechanical strain is converted into electrical output using strain gauges mounted on the elastic body. Such pressure transducers find application in high-pressure ranges and provide accurate and precise measurements.The work includes the evaluation of the performance characteristics of a hydraulic pressure transducer in the range of 800 MPa. After the 4-year study of the transducer, we have found it stable and accurate, with an uncertainty of less than 1%.

Keywords Pressure transducer · Metrology · Accuracy · Quality · Calibration factor · Stability

---

R. K. Sharma (✉) · A. Zafer · L. Kumar · J. Singh · S. Yadav · N. D. Sharma
CSIR-National Physical Laboratory, New Delhi, India
e-mail: sharmars2@nplindia.org

S. Yadav et al. (eds.), *Advances in Metrology*, Lecture Notes in Mechanical Engineering,
https://doi.org/10.1007/978-981-96-6418-4_8

Nomenclature [1]

m_i,	Mass of the weight (in kg)
ρ_{air}	Density of the air at temperature, barometric pressure, and humidity prevailing in the laboratory (in kg/m^3)
ρ_{mi}	Density of the ith weight (in kg/m^3)
γ	Surface tension of the pressure transmitting fluid (in N/m)
C	Circumference of the piston where it emerges from the fluid (in m)
A_0	Effective area of the piston-cylinder assembly at zero pressure (in m^2)
α_c & α_p	Thermal expansion coefficients of material of cylinder and piston respectively (in/°C)
T	Temperature of the piston-cylinder assembly (in °C)
T_r	Temperature at which A_0 is referred (in °C)
b	Pressure distortion coefficient of the effective area (in per Pa)
p_j	Jacket pressure during operation (in Pa)
p_z	Jacket pressure for which the clearance between the piston and cylinder is extrapolated to go to zero (in Pa)
D	Coefficient of the change of area with jacket pressure
Δ_p	Head correction in terms of pressure (in Pa)
H	The difference in height between the reference levels of the standard and gauge under test (GUT) (in m)
Δ_f	Density of the transmitting fluid (in kg/m^3)

1 Introduction

In pressure metrology, there are two fundamental pressure measurement methods: a liquid column manometer, and a piston manometer, also called dead weight tester (DWT) or pressure balance (PBs). Both are used for precise and accurate measurement of pressure as laboratories' primary standards. However, the liquid column manometer is restricted to the measurement of pressure nearly to the atmospheric pressure range due to the limitation of its column height. Thus, owing to the high accuracy and stability of PBs, they are mainly used as primary standards for high-pressure measurement and calibration at various National Metrological Laboratories and industries as reference pressure standards. The calibration is always done by comparing the measurements made by measuring equipment against a standard whose uncertainty is known and traceable to national/international standards. The PBs are calibrated using an internationally accepted cross-floating method [2, 3].

For industrial and shop floor level measurements, pressure transducers are used where there are two methods of pressure measurement: direct pressure measurement and indirect pressure measurement. In direct pressure measurement, a mechanical device undergoes a deformation in response to a pressure change, which is translated into an electric signal using a displacement transducer. In indirect pressure measurement, the mechanical deformation is too small to be measured, and therefore, indirect means are used to measure pressure with the help of another physical quantity, such as thermal conductivity.

The operational efficiency of a device under calibration is determined by the study of associated metrological parameters. Performance evaluation and stability studies have been performed on various pneumatic and hydraulic pressure gauges [4–7]. In the present study, we have analyzed the metrological behavior of a high-pressure industrial transducer by estimating its calibration factor (C_f), measurement uncertainty and stability over a period of time. It is important to check the consistency and stability of devices for their long-term usefulness. The major benefit of this study can further be utilized to improve the performance of gauges under calibration. It will help the users with equipment upgrades and quality control procedures. This study will also provide confidence to the end users to carry out internal calibrations.

2 Methodology and Experimental Setup

The pressure transducer used for the present investigation is a strain gauge type with a working range of (0–1000) MPa for hydraulic pressure with a measurement accuracy of 0.5% F.S. The calibration is carried out periodically using the Primary Hydraulic Pressure Standard to check its performance and stability. The experimental setup consists of (i) Primary Pressure Standard (make: Harwood) having working range up to and having measurement uncertainty associated with pressure measurements as 250×10^{-6} at $k = 2$, (ii) High-pressure transducer as gauge under test including a digital display unit for transducer output in volts, (iii) Digital multimeter for recording temperature of piston-cylinder assembly of primary standard via a PRT sensor, and (iv) Thermo hygrometer and barometric pressure transducer were used as supporting instruments for recording environmental conditions and, hence, for the calculation of air density. The experiment is performed at 8 pressure points by taking 6 readings at each pressure point in both the increasing and decreasing pressure cycles. To determine the gauge's performance under test in terms of calibration factor and stability, we need to repeat the experiment at regular intervals. We have performed the calibration in terms of pressure to calculate the calibration factor and associated uncertainty. Our primary standard has a 500 kg calibrated mass set having traceability from the mass standard.

The high-pressure transducer (HPT) under investigation, is connected to the pressure standard through mechanical fittings and has been tested in the (0–800) MPa range by performing its leak test. This industrial high-pressure transducer was calibrated against the national primary hydraulic pressure standard designated as NPL-H1 by direct pressure measurement method. The HPT was connected to the standard, and its sensor was placed on a rigid stainless steel table with a thickness of ~12 mm in order to avoid any vibrations during the calibration. During the process, the piston is rotated at (20–30) rpm with the help of an electric motor to reduce the

Fig. 1 Experimental setup for the testing and performance analysis

effect of friction. After assembling the test instrument with the standard instrument, a leak test is performed to ensure no leakage in the system. For the current study, the calibrated range of high-pressure transducer is (0–800) MPa. The complete pressure range is divided into 8 pressure points. Next; the experiment was carried out to check the performance and stability of the gauge under test. Figure 1 shows the setup for the testing and performance analysis.

3 Calibration Process and Experimental Investigation

The calibration room is maintained clean, and dust-free with no direct air-drafts. For calibration, necessary weights as per the requirement of the pressure point are loaded on the standard hanger assembly so that the desired pressure can be generated. The transducer output is recorded along with the mass value on the primary standard, the temperature of the piston-cylinder assembly as well as the environmental conditions. The observations were taken at pressures of 100, 200, 300, 400, 500, 600, 700, and 800 MPa with three readings in increasing cycles and three readings in decreasing cycles at each pressure point. The output of the transducer is recorded as a function of applied pressure, which is determined by the pressure standards.

Progressively, weights are loaded in the desired steps on the hanger assembly, and readings are taken at each point like the first reading. It is well known that the performance of a DWT is affected by various factors such as elastic distortion of the p–c assembly, temperature of the p–c assembly, effect of gravity on the masses, effect of buoyancy on the atmosphere upon the masses, hydraulic pressure gradients within the equipment, surface tension effects of the pressure transmitting liquids and head correction for the difference of height between the DWT and the pressure transducer.

The pressure of the test gauge is therefore, calculated by the pressure equation which is given by Eq. 1.

$$P_{DWT} = \frac{\sum_i m_i.g_{\mathrm{L}}(1 - \rho_{\mathrm{air}}/\rho_{mi}) + \gamma C}{A_0^P(1 + bP)[1 + (\alpha_C + \alpha_{\mathrm{P}})(T - T_{\mathrm{r}}]} \pm \Delta p \quad (1)$$

The observations were repeated six times at each pressure point. The transducer output is then least square fitted as a function of applied pressure. The calibration results thus obtained are discussed in the subsequent section.

The measurement uncertainty is evaluated based on the JCGM 100:2008 and NABL 141 guidelines [8, 9]. The standards measurement uncertainty associated u (y_c) with estimate y for a given value x by linear curve fitting is estimated by Eq. 2.

$$u(y_c) = \sqrt{\begin{matrix} \left[\left(\frac{\delta y}{\delta a_0}\right) \cdot u(a_0)\right]^2 + \left[\left(\frac{\delta y}{\delta a_1}\right) \cdot u(a_1)\right]^2 \\ +2\left[\left(\frac{\delta y}{\delta a_0}\right) \cdot \left(\frac{\delta y}{\delta a_1}\right) \cdot u(a_0) \cdot u(a_1) \cdot r(a_0, a_1)\right] + \left[\sigma(y_c)\right]^2 \end{matrix}} \quad (2)$$

where constants $a_{0,}$ $a1$, and $\sigma(a_0)$, $\sigma(a_1)$ are the standard deviations of the constants a_0 and a_1, respectively. ($\delta y/\delta a_0$) and ($\delta y/\delta a_1$) are the sensitivity coefficients of a_0 and a_1.

Finally the combined uncertainty associated with effective area measurements is then estimated by Eq. 3.

$$u_c^2\left(A_{eff}\right) = \sum_{i=1}^{n}\left(\frac{\partial A_{eff}}{\partial x_i}\right)^2 u^2(x_i) \quad (3)$$

where $u\ (x_i)$ represents the $\mathrm{i_{th}}$ $\frac{\partial A_{eff}}{\partial x_i}$ uncertainty contribution and is the corresponding sensitivity coefficient derived from partial differentiation of Eq. 3.

Calibration Factor is given by

$$\overline{C_f} = \frac{\sum_i^n C_{f(i)}}{n}$$

where $C_{\mathrm{f(i)}}$ is the ith calibration factor computed as follows:

$$C_{f(i)} = \frac{P_{r(i)}}{P_{t(i)}}$$

where

$$P_{r(i)} = P_{DWT(i)}$$

is the $\mathrm{i_{th}}$ reference pressure measured by the standard DWT and

$$P_{t(i)} = P_{Test(i)}$$

is the i_{th} indicated pressure of the Test (Transducer under test).

4 Results and Discussion

As mentioned, the HPT has been calibrated in the pressure range from 0 to 800 MPa, once in a year during 2021–2024 using the Primary Pressure Standard as a reference pressure gauge. The calibration factor and measurement uncertainty have been calculated from the recorded experimental data. The result obtained for the calibration factor (C_f) versus Pressure has been plotted year-wise in Fig. 2. It can be seen from the figure that the C_f values vary from 0.9915 to 1.0136 for all the years and hence, the measured calibration factor is close to unity. This shows that the pressure measured by the HPT agrees well with the standard pressure value, and hence, HPT is reasonably accurate. However, it is also clear from the figure that for the years 2021 and 2022, the C_f values are close to unity and varying within ± 0.009, while for the years 2023 and 2024, there is relatively larger variation ranging from a low value of 0.9915 to the highest value of 1.0136 showing a variation of 0.0221 in magnitude. This indicates a marginal deterioration in performance in the years 2023 and 2024 as compared to previous years.

Further, the variation in GUT output in volts with standard pressure is plotted in Fig. 3 along with their corresponding fitting equations for all the years. It is clearly visible from Fig. that the transducer output varies linearly with the standard pressure. The results of regression data for each year are also shown as equations in the inset. It is again observed that there is a decrease in the slope as well as the intercept from 2021 to 2024.

As mentioned earlier, associated measurement uncertainty in pressure was also estimated at each pressure point, i.e., from 100 to 800 MPa as per the GUM and NABL guidelines [7, 8]. All the type A and type B components are considered and tabulated in Table 1. The expanded uncertainty at a coverage factor of 2 thus obtained

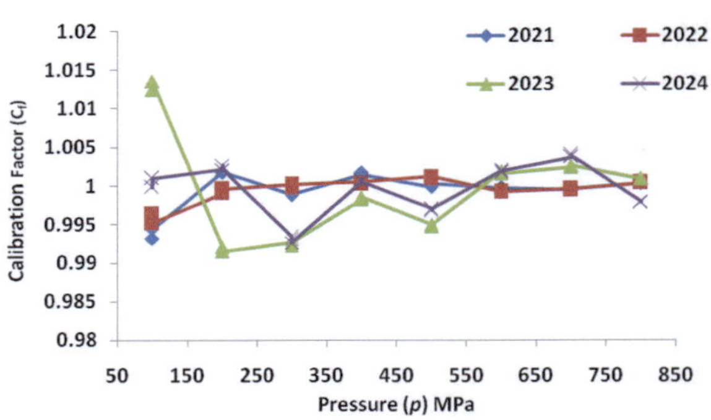

Fig. 2 Calibration factor versus pressure (MPa)

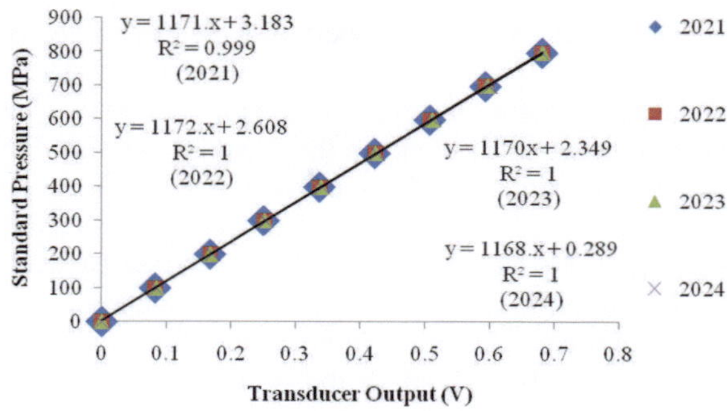

Fig. 3 Standard versus GUT pressure (MPa)

is plotted in Fig. 4. For all the years, it was observed that the highest measurement uncertainty is observed at the lowest measurement point, i.e., 100 MPa.

Further, the major contribution to the measurement uncertainty obtained is due to the uncertainty in regression or fitting coefficient, followed by uncertainty due to the zero setting and resolution in all the years. However, the repeatability is consistent and is quite low throughout the studied period. As the same hydraulic pressure standard is used to calibrate this HPT, the uncertainty is due to the standard used being the same for all the years, i.e., 125×10^{-6}x p.

Table 1 Summary of various parameters of uncertainty over the years

Uncertainty components	Year			
	2021	2022	2023	2024
Repeatability (Type A, MPa)	0.0200	0.0198	0.0247	0.0200
Regression (Type A, MPa)	0.2274	0.1896	0.4440	0.3772
Standard's uncertainty (Type B, MPa)	0.0125	0.0125	0.0125	0.0125
Zero setting (Type B, MPa)	0.0699	0.0345	0.000	0.0337
Resolution (Type B, MPa)	0.0435	0.0342	0.0348	0.0336
Combined uncertainty (u_c in % of reading)	0.24	0.20	0.45	0.38
Expanded uncertainty (U_{exp}) ($k = 2$, in % of reading)	0.49	0.39	0.89	0.76

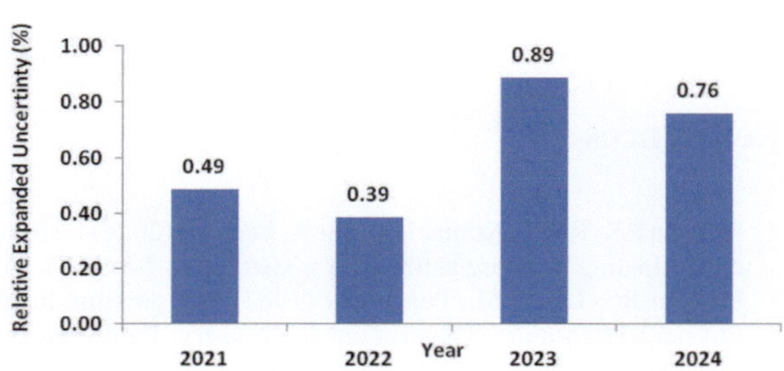

Fig. 4 Relative expanded uncertainty over the years

The lowest expanded uncertainty, i.e., ±0.39%, is observed in 2022, while the highest, i.e., ±0.89%, in 2023 as shown in Fig. 4. The increase in uncertainty in 2023 and 2024 is again due to the regression factors that show that transducer output has an increasing deviation from the standard pressure. This may occur due to a drift in the readings due to electronic components, which can be ascertained by analyzing more calibration data. Therefore, it can be concluded that the increase in uncertainty over the years has majorly resulted in increased scatter in the transducer readings, leading to an increase in C_f as well as uncertainty contribution due to regression fitting of the GUT data with the standard pressure.

However, throughout all the years, the measurement uncertainty has been well within the specified limit of ±1% at $k = 2$. Therefore, it can be concluded that the transducer is consistent over the high-pressure range and is accurate and stable over a studied period of 4 years although a minor deterioriation in the data has been observed. However, more calibration data would be required to deduce its long-term stability and gain more confidence, which we would take over the next few years in our future endeavors.

5 Conclusions

The variation in the performance of a high-pressure transducer was investigated over a four-year period. It was observed that although the transducer showed good behavior over a period of 4 years, a minor increase in C_f as well as uncertainty contribution due to regression fit was found. The transducer behavior showed excellent repeatability for each measurement. It was observed that the pressure transducer showed reproducibility of less than 1% which is well within the acceptable limit for high pressure applications. The HPT also has good accuracy. Hence, for the 4 years, the HPT has exhibited good performance for high pressure range with slight deterioration in the years 2023 and 2024. Further study of the metrological characteristics will be done for long-term stability.

Acknowledgements The authors are thankful to the Director CSIR-NPL for his continuous support and motivation.

References

1. Sharma RK, Rab S, Kumar L, Zafer A, Yadav S (2021) Design and development of an indigenous cross-floating pressure calibration system up to 140 MPa. MAPAN 36:295–303
2. Dadson RS, Lewis SL, Peggs GN (1982) The pressure balance: theory and practice. National Physical Laboratory, Department of Industry, Published by her Majesty's Stationary Ofiice, London
3. Heyemanand PLM, Welch BE (1975) Experiment thermodynamics (vol 2, p 147). In: Le Neindre B, Vodor B (eds) Butterworths, London

4. Kajikawa H, Kobata T (2019) Different long-term characteristics of hydraulic pressure gauges under constant pressure applications. ACTA IMEKO 8(3):19
5. Kojima M, Kobata T, Fujii K (2015) Long-term stability and zero drift of digital barometric pressure gauges. Metrologia 52:262–266
6. Yadav S et al (2006) Studies on the stabilities of various types of industrial pressure measuring devices. J Sci Ind Res 65:721–724
7. Singh J, Kumar A, Sharma ND, Bandyopadhyay AK (2011) Reliability and long-term stability of a digital pressure gauge (DPG) used as a standard—a case study. MAPAN 26(2):115–124
8. Guidelines for estimation and expression of uncertainty in measurement. NABL 141
9. BIPM I et al (2008) Evaluation of measurement data—guide for theexpression of uncertainty in measurement. JCGM 100

Straightness Measurement of Long Cylindrical Guide Rod Using Autocollimator with Aid of Precision Level

H. S. Chethan, Khushboo, and K. Niranjan Reddy

Abstract Long cylindrical guide rods are widely used in large industries such as turbine power systems, defense equipments to guide bowls. These cylindrical guide rods (having lengths in meters) require precision form control like straightness. This straightness parameter can't be measured using conventional techniques where longer datum is unavailable or using angle measuring devices directly. This paper aims to evolve a new measurement method for checking the straightness of a long cylindrical guide rods by an autocollimator with the aid of the Precision level. This newly developed method is validated by comparing the straightness results of precision cylinder (length of 500 mm) using the conventional method.

Keywords Straightness · Large cylindrical guide rod · Autocollimator with precision level

1 Introduction

Long cylindrical rods are key components in various industrial applications, where precision engineering is crucial for performance and reliability. Cylindrical rods are being used in machines and plays a vital role in the functions of machine tools. The main functions of cylindrical rods are it acts as (i) structural members to support the machine elements, (ii) a guide for axial movements for parts of the machine, where parts will travel to and forth over the guide (e.g.: mixer machine), (iii) a column for axial and rotary movements of parts of the machine, where parts will rotate and move up and down over the column (e.g.: radial drilling machine). To carry out the functionalities mentioned in (ii) and (iii), cylindrical rods should have good form (geometry) and accurate dimensions. One of the most critical form parameters for these rods is straightness. If the straightness of the cylindrical rod is not within the designed specification, it may cause many unpredictable problems like vibration,

H. S. Chethan (✉) · Khushboo · K. Niranjan Reddy
Central Manufacturing Technology Institute, Bengaluru, India
e-mail: chethan@cmti.res.in

S. Yadav et al. (eds.), *Advances in Metrology*, Lecture Notes in Mechanical Engineering,
https://doi.org/10.1007/978-981-96-6418-4_9

friction, wear & tear, etc. Hence, ensuring that a rod is straight within specified tolerances is crucial for its proper functioning in high-precision industries like turbine manufacturing, aerospace, and defense. These industries demand components that meet strict quality and precision standards to ensure safety, reliability, and efficiency in their applications. Therefore, measurement and ensuring straightness are important and critical tasks to confirm the quality and functionality of long cylindrical rods. Traditional methods for measuring the straightness of cylindrical rods typically involve the use of surface plates and dial gauges. However, these methods encounter significant challenges when applied to long cylindrical rods that are heavy and extend over several meters [1]. The primary difficulties include handling the rod's substantial weight and accurately identifying the same high points along its length. To address these limitations, this paper presents an innovative method for measuring the straightness of long cylindrical rods using an autocollimator with the aid of a precision level. An autocollimator is a non-contact angle measuring instrument using the geometric light reflection principle, which measures small angular displacements with high precision, making it suitable for measuring straightness over long distances [2, 3]. The precision level is a small angle measuring instrument, which is mainly used to measure the inclination of the relative horizontal plane [4] and is capable of providing accurate leveling references, ensuring that readings are taken at the same high point.

To validate this method, conducted experiments comparing the straightness measurements obtained from a precision cylinder (a shorter cylindrical rod) using both the conventional method, i.e., surface plate with dial gauge method and the new autocollimator with the precision level method, and detailed the experimental setup, measurement procedures, and comparative analysis results.

Xinchun Chen et al. [1] developed a device and method for measuring the straightness of large piston rods with a high length-to-diameter ratio using a time-domain three-point technique. This method addresses the various requirements of straightness measurement during production. Additionally, the effects of three error sources—axial rotation angle, probe spacing, and sampling interval—on the measurement results were thoroughly discussed.

W. Gao, J et al. [5] presented a new multi-probe method for accurately measuring the straightness profiles of cylinders. This scanning probe method is developed from the traditional three-probe method, which can measure the profile accurately without the influence of motion errors of the scanning stage. Two sets of three-probe method sensor units are used in the developed method so that the zero-adjustment errors between the probes can be evaluated.

Krzysztof Stepien et al. [6] investigated methods for accurate in situ cylindricity measurements. These methods are categorized into two main groups: multisensor and V-block methods. They concluded by outlining the most significant advantages and disadvantages of each method.

In the existing literature, there is a noticeable gap regarding the on-site straightness measurement of long cylindrical guide rods using an autocollimator. The focus has instead been on other methods or on developing measuring instruments that require moving the guide rod to a laboratory. In this present work, developed a new method for on-site straightness measurement of long cylindrical guide rod using an autocollimator with the help of precision level.

2 Problem Identification and Solution

Straightness measurement of a cylindrical guide rod by autocollimator requires obtaining measurement readings at the same high point over the cylindrical surface. Maintaining the same high point while capturing readings by autocollimator is not possible by using existing methods. To address this challenge, a specialized fixture consisting of a V-guide with a plate (Fig. 1a) has been designed to accommodate the precision level to maintain the same high point and plane mirror of autocollimator to capture the straightness readings. This fixture with V-guide securely positions over the surface of the cylindrical rod aligned along its longitudinal axis during measurement, restraining any lateral movement that could introduce measurement errors. The plate provides a stable base for precision level and plane mirror of autocollimator. Figure 1b shows the assembled condition of fixture.

The autocollimator emits a collimated light beam that strikes the plane mirror and is reflected back to the autocollimator. Any deviation in the mirror's angle causes a shift in the reflected beam's position on the autocollimator's sensor. This positional shift is measured as an angular deviation, which is then converted into a straightness measurement.

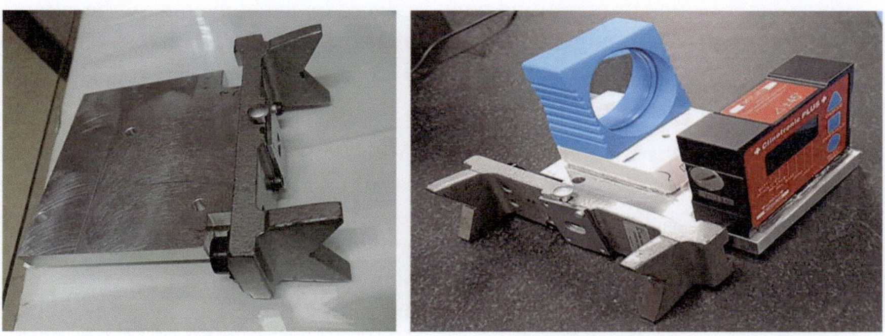

(a) (b)

Fig. 1 **a** Fixture, **b** Fixture in assembled condition

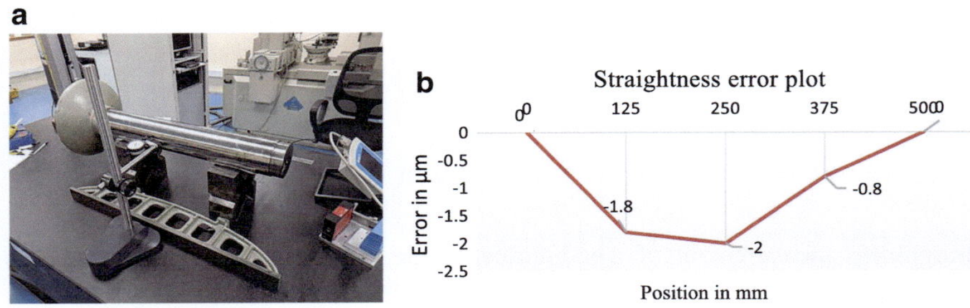

Fig. 2 Surface plate with dial gauge method

3 Experimentation Methods

To validate the method straightness of the precision cylinder having 500 mm length is measured by conventional surface plate with dial gauge method and autocollimator with precision level method; both methods are explained briefly, and results are presented.

(a) **Conventional Surface Plate with Dial Gauge Method**: The surface plate with dial gauge method is a traditional technique for checking straightness. It uses a surface plate and/or a straight edge as references, and a dial gauge to measure straightness as shown in Fig. 2. In this method, the straight edge and the component are placed on the surface plate. The dial gauge is set against the straight edge as a reference, and its measuring point is placed on the high point of the component's surface. The straight edge with dial gauge is aligned with the component and straightness deviations are measured along the length of the rod. Readings are shown in Fig. 2.

(b) **Autocollimator with Precision Level Method**: Autocollimator is an optical instrument used to measure small angular differences. Autocollimator is very sensitive to measure the change in angles and provides very accurate measurements. The setup for the autocollimator includes placing the autocollimator outside the measurement axis and attaching a plane mirror and precision level to the fixture, with the autocollimator and mirror aligned to the measurement axis, as shown in Fig. 3. Readings are taken at each interval by maintaining the same high point with the help of a precision level. Readings are shown in Fig. 3.

4 Results and Discussions

The results from both methods are presented in Table 1. The En (Normalized Error) ratio, which compares the measurement uncertainty and error between the two methods, was found to be less than 1. This indicates that the new method is validated.

En Ratio Formula

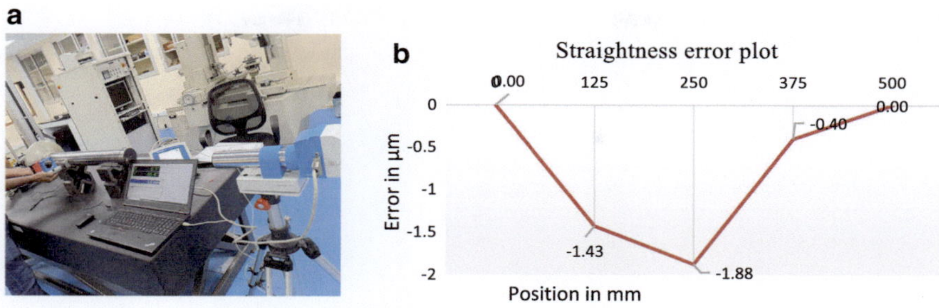

Fig. 3 Autocollimator with precision level method

Table 1 Results and En ratio

Parameter	Autocollimator with precision level method (Aut) (μm)	Dial gauge method (Ref) (μm)	En ratio
Straightness (x)	1.88	2.00	0.26
Uncertainty (U)	± 0.3	± 0.44	

$$\boldsymbol{En} = \frac{x_{Aut} - x_{Ref}}{\sqrt{U_{Aut}^{2} + U_{Ref}^{2}}}$$

where, x_{Aut} = Straightness error by autocollimator method.

x_{Ref} = Straightness error by dial gauge method.

U_{Aut} = Uncertainty of measurement by autocollimator method.

U_{Ref} = Uncertainty of measurement by dial gauge method.

On validation, the straightness measurement of a long cylindrical guide rod with a length of 3125 mm was conducted using the autocollimator with precision level method. The measurement setup for this long cylindrical rod is shown in Fig. 4. The autocollimator with fixture was aligned to the long cylindrical guide rod, and straightness measurements were taken along the entire length of the rod. Figure 4 shows the results obtained, which indicates a straightness value of 36.9 μm.

Fig. 4 Measurement setup for long cylindrical rod

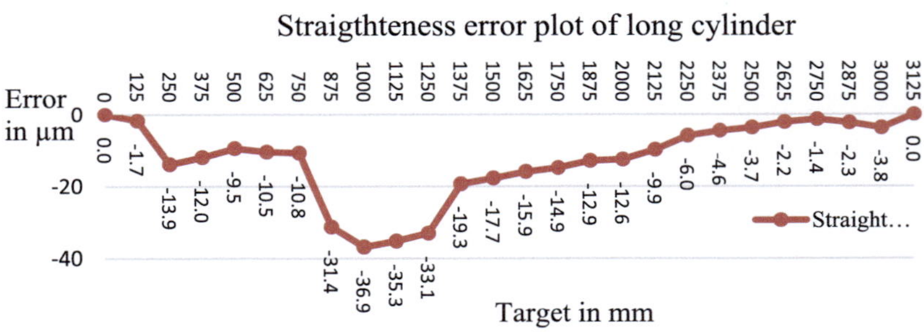

Fig. 5 Straightness error plot of long cylindrical rod

5 Conclusion

A new method autocollimator with a precision level for straightness measurement of long cylindrical guide rods has been developed. This method is validated by comparing the straightness results of a precision cylinder using both the conventional method and this new autocollimator method.

This new method successfully addresses the challenge of measuring both horizontal and vertical straightness of long cylindrical guide rods, enabling precise on-site straightness measurements.

The fixture design can be improved to make it more comfortable and easier for metrologists to use. This can be taken up in future research.

Acknowledgements The authors are thankful to Mr. Nagahanumaiah, CMTI, for providing the opportunity and to metrology lab personnel for support during the experiments.

References

1. Chen X et al (2019) Study on measuring straightness of piston rod in large oil cylinders. J Phys Conf Ser 1303:012137
2. Vaishnani AR (2019) Measurement of straightness error of machine guideways using autocollimator. © 2019 JETIR 6(2) (ISSN-2349–5162)
3. Kaewpho S et al (2023) An application of autocollimator for strip surface profile measurement. J Phys Conf Ser 2431:012022
4. Yang Y et al (2022) Research on straightness calibration method of straight edge working face based on electronic level. MATEC Web Conf 363:01013
5. Gao W, Kiyono S (1997) On-machine roundness measurement of cylindrical workpieces by the combined three-point method. Measurement 21(4):147–156
6. Krzysztof Stepien (2009) Research on accurate in situ measurements of cylindricity. Fundam Appl Metrol Lisbon, Portugal

Non-visual Effect of Color-Tunable Smart LED Sources

Vijeta, Saurabh Babu Saxena, Shibu Saha, V. K. Jaiswal, and Parag Sharma

Abstract Color is a ubiquitous aspect of our daily lives, present in everything we encounter. This paper focuses on analyzing smart lighting systems with tunable colors, examines the spectral power distribution of these lights to investigate their non-visual effects, specifically evaluating the circadian action factor (CAF) value.

Keywords Smart Tunable LED · Circadian action factor (CAF) · Sphere Spectroradiometer system

1 Introduction

Light plays a pivotal role in governing the human circadian system, exerting substantial influence over a range of physiological and psychological processes [1]. Consequently, there has been a notable surge of research interest in optimizing light sources to enhance circadian performance. The investigation of the non-visual attributes of artificial lighting, in addition to its visual and color characteristics, holds significant importance in the current era of smart lighting [2]. Night shift workers often face challenges due to the misalignment between their work schedule and the body's circadian rhythm as prolonged exposure to artificial lighting suppresses melatonin. Therefore, understanding the non-visual circadian effects, quantified through the assessment of the circadian action factor (CAF), is crucial for human health and work performance in various tasks [3, 4].

This paper explores the use of a smart tunable LED source that can be adjusted to different color temperatures and colors, specifically designed to provide tunable circadian lighting performance. The non-visual effect of tunable colors of smart

Vijeta · S. B. Saxena · S. Saha · V. K. Jaiswal · P. Sharma (✉)
CSIR-National Physical Laboratory, New Delhi, India
e-mail: sharmap2.npl@nic.in

Academy of Scientific and Innovative Research (AcSIR), Ghaziabad, India

Vijeta
Physikalisch - Technische Bundesanstalt (PTB), Braunschweig, Germany

S. Yadav et al. (eds.), *Advances in Metrology*, Lecture Notes in Mechanical Engineering,
https://doi.org/10.1007/978-981-96-6418-4_10

lighting can be determined by the CAF values corresponding to their spectral power distribution (SPD).

2 Theory

Numerous studies have been conducted to determine the energy efficiency and color rendering parameters such as luminous efficacy of radiation (*LER*), luminous efficacy (*LE*), color rendering index (*CRI*), and correlated color temperature (*CCT*) of light sources [5, 6]. With the advent of photobiological research and LED technology, it is now possible to optimize the spectrum of the light source for the circadian effect. It is important to parameterize the effect of artificial lighting on human health and work performance [7, 8]. These parameters depend on the SPD of a light source, which can be used to determine circadian and visual effects of lighting. Parameters illustrating the visual and non-visual impacts of any light source can be evaluated using *SPD* [4, 9].

The luminous efficacy of a light source (*LE*), (lm/W) measures the ability of a light source to emit visible light using a given amount of power, i.e., in lumens per watt. Also, the luminous efficacy of radiation (*LER*), corresponding to the visual impact of a light source can be expressed as

$$LER\left(\frac{lm}{W}\right) = \frac{K_o \int_{380nm}^{780nm} V(\lambda)S(\lambda)d\lambda}{\int_{380nm}^{780nm} S(\lambda)d\lambda} \tag{1}$$

To quantify the circadian effect of the light spectrum, the circadian efficacy of radiation (*CER*) is evaluated as [4, 9].

$$CER\left(\frac{blm}{W}\right) = \frac{K_{co} \int_{380nm}^{780nm} C(\lambda)S(\lambda)d\lambda}{\int_{380nm}^{780nm} S(\lambda)d\lambda} \tag{2}$$

The Luminous Efficacy of Radiation (*LER*) and Circadian Efficacy of Radiation (*CER*), Eqs. 1 and 2, can be defined as the ratio of the luminous flux to the radiant flux ($S(\lambda)$) and the ratio of the circadian flux to the radiant flux ($S(\lambda)$), respectively.

The parameters, *LER* and *CER,* are also defined as optical parameters explaining how well/brightly the emitted radiation (visually and non-visually) from any light source is perceived by the vision and non-vision system of the average human eye, respectively. In the formulation of *LER* and *CER*, $S(\lambda)$ indicates the SPD of the light source and $V(\lambda)$, and $C(\lambda)$ represent the normalized photopic and circadian spectral luminous efficiency function respectively as shown in Fig. 1 [4].

Despite several reported circadian spectral sensitivity functions, C(λ), and their associated limitations, the model proposed by Gall et al. has been selected for the calculation of circadian effect-related parameters [10, 11]. The $V(\lambda)$ has an absorption maximum at 555 nm (green), whereas $C(\lambda)$ [11] has an absorption maximum at

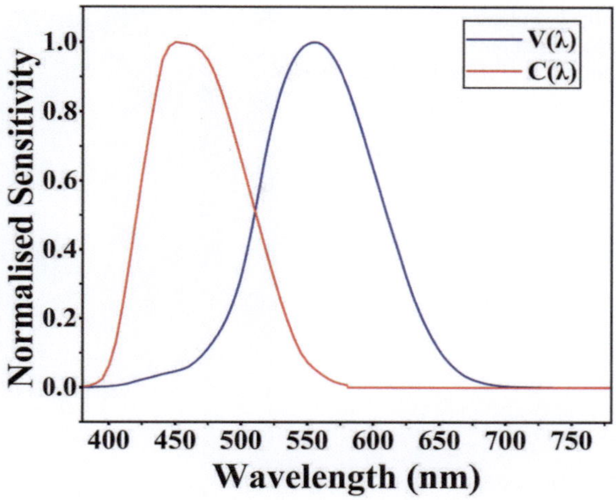

Fig. 1 Normalized circadian spectral sensitivity function $C(\lambda)$, and photopic spectral sensitivity functions $V(\lambda)$ [2, 10, 11]

460 nm (blue), determined from data of the melatonin suppression action spectrum. Further, K_o (683 lm/W) is maximal spectral luminous efficacy for vision and K_{co} (683 blm/W) is a maximum circadian efficacy value for the non-visual system, respectively [2]. Circadian lumen (cirlm) or biolumen (blm) are the proposed units for measuring circadian flux, hence, *CER* and K_{co} units are either biolumen per watt (blm/W) or cirlm/W.

The non-visual effects of light sources on humans can be characterized by assessing *CAF* value [2, 12]. Further, *CAF* value can be optimized for circadian performance because of its positive correlation with melatonin suppression, defined as the ratio of *CER* to *LER* values [9, 12, 13].

$$CAF\left(\frac{blm}{lm}\right) = \frac{CER(\text{blm/W})}{LER(\text{lm/W})} \tag{3}$$

The circadian action factor (CAF), Eq. (3), also represents the biological action per unit of visual response. The evaluated CAF value is an important factor for eliciting the effect of artificial lighting on human health or the biological clock. The variation of CAF value is dependent on the exposure and intensity of an artificial light source. CAF values vary during the 24-h day and night cycle as high CAF values are observed during daylight and low CAF values correspond to the night time.

3 Experimental Setup

For this investigation, a commercially available 10W wifi enabled smart tunable LED source adapted to different color temperatures and colors was used as test source. A sphere spectroradiometer system of diameter 2 m (as shown in Fig. 2) was used

Fig. 2 Sphere spectroradiometer system present at CSIR-NPL, India

for the measurement of SPD corresponding to different colors, produced by smart tunable LED source.

The sphere spectroradiometer system employed was calibrated with a spectral radiant flux (SRF) source traceable to the National Institute of Standards and Technology (NIST), USA, through NAVLAP certification. All optical measurements were conducted at an ambient temperature of (25 ± 2) °C under steady-state thermal and electrical conditions.

4 Results and Discussions

The smart LED source was tuned to different colors, C1–C28 and corresponding SPDs measured by sphere spectroradiometer system as shown in Fig. 3. The measured SPDs were characterized to evaluate their visual performance, non-visual effects, and color quality.

The evaluated chromaticity coordinates (x, y) of the different colors generated by optimizing the SPDs are shown by stars embedded in the CIE 1931 chromaticity diagram in Fig. 4.

Further, the CAF values corresponding to the tunable colors have been evaluated (using Eq. 3) from the measured SPDs.

The variation of CAF in correspondence to the chromaticity coordinates of tunable colors embedded in the CIE 1931 chromaticity curve is shown in Fig. 5.

It is considered that the variation of CAF is continuous within these discrete points. It is evident from Fig. 5 that the color (C7) with chromaticity coordinate (0.182, 0.26) results in a higher CAF value (1.81 blm/lm) with a CRI of 46.5, which is not well suited for circadian performance. Tunable orange color (C25) with chromaticity coordinates (0.520, 0.418) with a CRI of 21.9 is in favor of the well-being of the circadian system and may be used to improve response time in shift workers as it corresponds to a lower CAF value of 0.29 blm/lm. Tunable red color (C28) with chromaticity coordinates (0.669, 0.315) and CRI of 55 has the lowest CAF value of

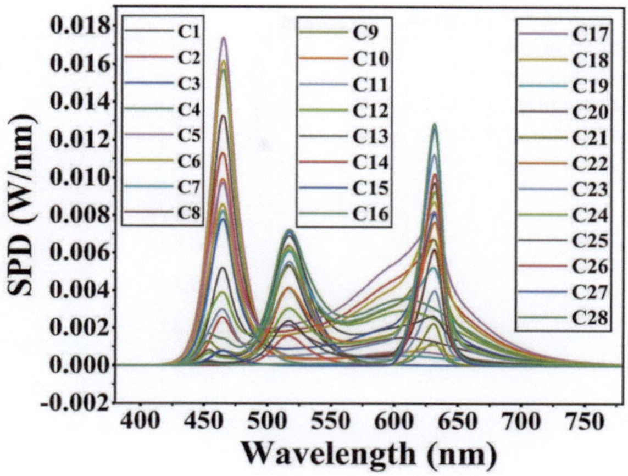

Fig. 3 SPDs of different colors generated by smart tunable LED source

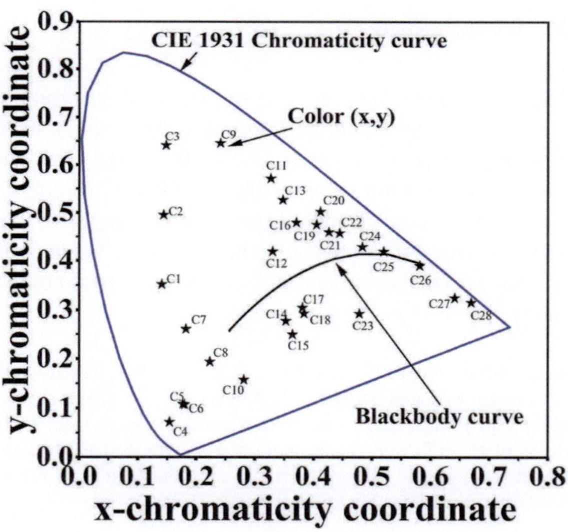

Fig. 4 Chromaticity coordinates of different colors generated by tunable LED source, embedded in CIE 1931 chromaticity diagram

0.04 blm/lm, also supports the fact of the impact of red-light exposure, particularly during night-time, on alertness and performance metrics [14, 15].

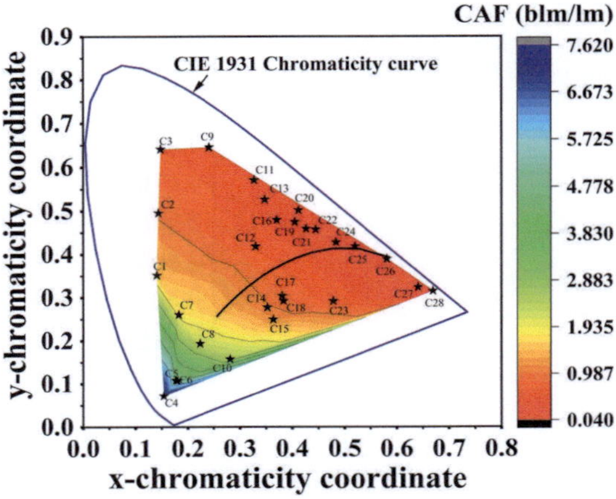

Fig. 5 CAF values of different colors (C1–C28) generated by tunable LED source plotted in CIE 1931 chromaticity diagram

5 Conclusion

This study identifies red and orange as the optimal tunable colors for supporting circadian rhythms and alertness at night-time. However, it is important to note that there may be risk to human health due to the considerable variation in CAF values caused by instantaneous tunability. This work could have significant implications for a wide range of applications for late night or shift workers for synchronizing their circadian rhythm. It also offers a multifaceted approach that can be tailored to specific spectral optimizations corresponding to color choices for relaxing mood and sleep cycles.

Acknowledgements The authors thank Director, CSIR–National Physical Laboratory for the encouragement and support. Technical support from the members of the Optical Radiation Metrology group is acknowledged. Financial support from the Bureau of Energy Efficiency (BEE), the Ministry of Power, Government of India, for the LED testing and calibration facility is acknowledged. One of the authors, Vijeta is also thankful to University Grant Commission (UGC) for providing the fellowship under the UGC-SRF scheme.

References

1. LeGates TA, Fernandez DC, Hattar S (2014) Light as a central modulator of circadian rhythms, sleep and affect. Nat Rev Neurosci 15(7):443–454. https://doi.org/10.1038/nrn3743
2. Zheng L-L et al (2016) Spectral optimization of three-primary LEDs by considering the circadian action factor. IEEE Photonics J 8(6):1–9. https://doi.org/10.1109/JPHOT.2016.2623667
3. Figueiro MG, Pedler D (2020) Red light: a novel, non-pharmacological intervention to promote alertness in shift workers. J Safety Res 74:169–177. https://doi.org/10.1016/j.jsr.2020.06.003

4. Dai Q, Shan Q, Lam H, Hao L, Lin Y, Cui Z (2016) Circadian-effect engineering of solid-state lighting spectra for beneficial and tunable lighting. Opt Express 24(18):20049–20059. https://doi.org/10.1364/OE.24.020049
5. Vijeta, Manchalwar N, Saha S, Jaiswal VK, Sharma P (2023) Traceability of total spectral radiant flux (TSRF) scale using spectral irradiance and total luminous flux scale at CSIR-NPL, India. MAPAN 38(3): 629–640. https://doi.org/10.1007/s12647-023-00661-x
6. Vijeta, Manchalwar N, Saha S, Jaiswal VK, Sharma P (2024) Suitability of commercially available array spectroradiometer for sphere spectroradiometric applications. J Opt 53(2): 968–979. https://doi.org/10.1007/s12596-023-01299-3
7. Roberts JE (2010) Circadian rhythm and human health. Dep Nat Sci Room 813:2010
8. Stevens RG (2006) Artificial lighting in the industrialized world: circadian disruption and breast cancer. Cancer Causes Control 17(4):501–507. https://doi.org/10.1007/s10552-005-9001-x
9. Oh JH, Yoo H, Park HK, Do YR (2015) Analysis of circadian properties and healthy levels of blue light from smartphones at night. Sci Rep 5(1):11325. https://doi.org/10.1038/srep11325
10. Gall D, Lapuente V (2002) Beleuchtungsrelevante Aspekte bei der Auswahl eines förderlichen Lampenspektrums. Licht 54(7):860–871
11. Gall D, Bieske K (2004) Definition and measurement of circadian radiometric quantities. Univ.-Bibliothek
12. Vijeta, Ray B, Saha S, Jaiswal VK, Sharma P (2024) Investigation on spectral profiles of white light generated by a smart tuneable LED source at different CCTs for optimized circadian performance. J Mod Opt 71(16–18):571–580. https://doi.org/10.1080/09500340.2024.2440522
13. Bellia L, Bisegna F, Spada G (2011) Lighting in indoor environments: visual and non-visual effects of light sources with different spectral power distributions. Build Environ 46(10):1984–1992. https://doi.org/10.1016/j.buildenv.2011.04.007
14. Figueiro MG, Sahin L, Wood B, Plitnick B (2016) Light at night and measures of alertness and performance: implications for shift workers. Biol Res Nurs 18(1):90–100. https://doi.org/10.1177/1099800415572873
15. Canazei M, Pohl W, Bliem HR, Weiss EM (2017) Acute effects of different light spectra on simulated night-shift work without circadian alignment. Chronobiol Int 34(3):303–317. https://doi.org/10.1080/07420528.2016.1222414

Studies on Pressure Dependent Dielectric Properties of Lead Magnesium Niobate-Lead Titanate-Lead Magnesium Tungstate Relaxor

Vinay Kumar

Abstract Precise pressure monitoring is essential in many manufacturing and processing sectors. A single instrument cannot be used if the required pressure range is too wide. Additionally, pressure measuring instruments may be sensitive for small pressure changes and insensitive to temperature over a range of operating temperatures. Excellent electro-mechanical properties of single crystals and ceramics are guaranteed by the well-known chemical inhomogeneity of relaxor-based solid solutions, which makes such materials potentially appealing for the development of high-efficiency sensors. The Curie-range temperature insensitivity of relaxor ferro-electric can be exploited to create devices whose output is not affected by temperature fluctuations. In the present investigation author studied the temperature and pressure dependent dielectric properties of Lead Magnesium Niobate-Lead Titanate-Lead Magnesium Tungstate Relaxor (PMN-PT-PMW) system by varying the composition of PT and excess Tungsten Oxide (WO3). This study leads to use of PMN-PT-PMW relaxor for high pressure sensor up to 415 MPa.

Keywords Pressure · Relaxor · PMN-PT-PMW · Dielectric · Sensor

1 Introduction

Precise pressure monitoring is essential in many manufacturing and processing sectors, including 'steel', 'aerospace', 'automotive', 'high-strength material synthesis', and 'research', etc. Pressure is one of the most crucial physical properties that requires precise measurement in the industry because of safety regulations as well as quality concerns. Pressure is an important physical quantity used in many applications, including 'water supply', 'pipeline construction', 'hydraulic lifts', 'hydraulic presses', 'hydraulic diggers', 'thermal power', 'hydraulic power', 'aviation of flights, 'flow meters', to propel aircraft, drinking with a straw, to inflate

V. Kumar (✉)
Jubail Industrial College, Jubail, Saudi Arabia
e-mail: vinayvarsh@gmail.com

S. Yadav et al. (eds.), *Advances in Metrology*, Lecture Notes in Mechanical Engineering,
https://doi.org/10.1007/978-981-96-6418-4_11

toy balloon and maintaining ideal pressure in tires, tubes, and human blood, etc., among other quantities. The choice of instrument to measure pressure depends upon range of interest and required accuracy. A single instrument cannot be used if the required pressure range is too wide. Additionally, pressure measuring instruments may be sensitive for small pressure changes and insensitive to temperature over a range of operating temperatures. 'Manometers Pressure Gauges', 'Bourdon Tube Pressure Gauges', 'Diaphragm Pressure Gauges', 'Dead Weight Pressure Gauges', 'MEMS Pressure Gauges', 'Manganin Pressure Gauges', 'Bellows Pressure Gauges', 'Piezometer Pressure Gauges', and so forth are a few examples of pressure measurement instruments [1].

As a result of ongoing material and technological advancements since the discovery of ferroelectricity in single crystal materials (Rochelle salts) in 1921 [2] and its subsequent extension to poly-crystalline materials (Barium Titanate, BaTiO3) in the 1940 [3], large number of materials has been discovered. Due to fundamental scientific interest and potential technological applications, extensive research activities have been carried out in the field of ferroelectric relaxors during the past five decades [4]. As a result, many compounds have been discovered to understand relaxor mechanism, to use relaxors in different applications, and research is still ongoing. Ferroelectric relaxors Lead Magnesium Niobate [PMN], Lead Zinc Niobate [PZN], Lead Scandium Niobate [PSN], Lead Scandium Tantalate [PST], Lead Iron Tungstate [PFW], and Lead Indium Niobate [PIN] [5–10] are few examples of relaxors used in a wide range of applications [11–18]. Few publications [19–23] demonstrate that this material can be used also as pressure sensor.

Ferroelectric relaxor materials are characterized by a diffusive phase transition and exhibit high electrostriction [24, 25]. Various lead-based compounds with perovskite structure (ABO3) (A cations on the corner, B cations on body center positions, and oxygen anions occupy the faces and form octahedral surrounding the B-site) exhibit relaxor ferroelectric behavior. The Curie-range temperature insensitivity of relaxor ferroelectric can be exploited to create devices whose output is not affected by temperature fluctuations. Excellent electro-mechanical properties of single crystals and ceramics are guaranteed by the well-known chemical inhomogeneity of relaxor-based solid solutions, which makes such materials potentially appealing for the development of high-efficiency sensors. Among the compounds, PMN has been widely studied either single or in combination with other materials [26–32] since its discovery due to its high dielectric constant and difficulty in obtaining perovskite structure [33].

Pb(Mg1/3Nb2/3)O3–xPbTiO3 (PMN–PT) system that are close to the rhombohedral-tetragonal morphotropic phase boundary [34] is of particular interest. Micro inhomogeneities can make it difficult to create ceramics with consistent qualities. Furthermore, the piezoelectric characteristics of PMN-based ceramics and single crystals are influenced by their relaxor behavior. The presence of cations in various charge states in the octahedral site, which are responsible for the generation of polar nanoscale areas relaxing in the non-polar matrix, is essential for the development of a relaxor state [35]. There is enough evidence to suggest that the B cations of the perovskite phase are organized within the nanoscale domains, although the nature of the polar areas is still a topic of much debate. Adding Pb(Mg1/2W1/2)O3 perovskite, which contains ordered magnesium and tungsten ions, could therefore be expected to enhance the amount of ordered nanoscale domains in PMN-PT ceramics, hence altering their electrical characteristics. In this context, the present study aims to investigate the pressure effect on Lead Magnesium Niobate-Lead Titanate-Lead Magnesium Tungstate (PMN-PT-PMW) ferroelectric relaxor material. This study should lead to the first-time development of capacitive pressure sensor materials using PMN-PT-PMW relaxor for measuring high hydrostatic pressures and almost independent of temperature fluctuations.

2 Experimental

Lead oxide is lost during sintering in lead-based ferroelectrics, which eventually lowers the ferroelectric properties of ceramics and encourages the production of low-dielectric-permittivity pyrochlore phases [36]. To get over this problem, the specimen of a tertiary system PMN-PT-PMW was created by adjusting the excess Tungsten Oxide (WO3) and PT concentrations and using the columbite precursor technique.

2.1 Specimen Preparation

Two distinct batches of specimens were created utilizing various compositions. PT composition was varied in one batch in the system: (1-x–y)PMN-xPT-yPMW + ZWO3, where x = 10, 27, 35%, y = 30%, and z = 5%. The composition of $_{\mathrm{WO3}}$ was altered in a different batch, with z = 0, 3, and 5%, in the system 43%PMN-27%PT-30%PMW + zWO3. Using the columbite precursor technique [37], specimens were prepared using high purity oxides (Lead Oxide (PbO)—99.9%, Magnesium Oxide (MgO)—97%, Niobium Oxide (Nb2O5)—99.9%, Titanium Oxide (TiO3)—99.9%, and WO3—99.99%). In the present study, all the oxides except PbO were taken stoichiometrically and ground in an agate mill in acetone medium for ten to twelve hours. After that, the mixture was calcined for two hours at 1273 K. After adding PbO to the calcined mixture, it was calcined once again for two hours at 1073 K. The calcined powder was compressed into 18 mm diameter and 1–2 mm thick pellets, after adding Poly Vinyl Alcohol (PVA) binder. After burning binder at 773 K for 30 min then pellets were sintered at 1423 K in a lead rich environment. Ultimately, the silver electrodes were formed on pellets through the vacuum evaporation process. Figure 1 depicts the flowchart for the specimen preparation procedure.

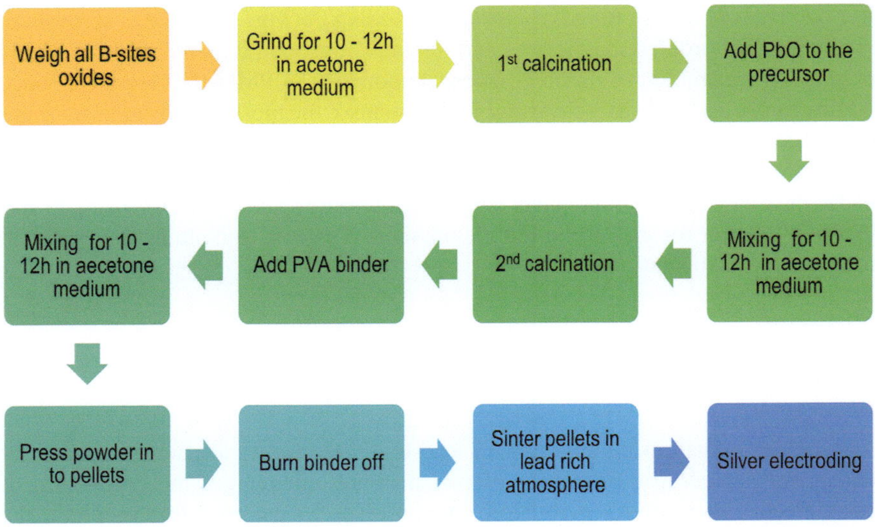

Fig. 1 Specimen preparation steps

2.2 Dielectric Measurement

Using an auto-ranging digital capacitance bridge (Andeen—Hagerling, Model 2500A, USA), capacitance and dielectric loss (*tan* δ) were measured at a fixed frequency of 1 kHz with a precision of 0.1 ppm throughout the capacitance range.

2.2.1 Dielectric Measurement at Various Temperatures

The design of the specimen holder used for the temperature and pressure measurement is shown in Fig. 2. This specimen holder made up of Invar is composed of five flanges, with a hole in the central flange that may hold up to two specimens with varying diameters (maximum 25 mm). There are two additional flanges on either side of the main flange; each one of them has a golden bellow for the electrical contact with the specimen. To measure the capacitance, the outer three BNC connectors were connected electrically with the specimens. One bellows is connected to central flange. The pressure is held in place by the outer flange. A thermostatically controlled temperature bath, model RTD 8dd Nes Lab, USA, was used to vary the temperature for the dielectric measurement. The specimen's temperature can be kept within 0.05 K with this bath. A digital temperature indicator with a precision of 0.01 K and a chromel-alumel thermocouple were used to measure the specimen's temperature. The temperature variation range that was selected was 283 to 323 K. A specimen holder was put in a copper vessel that was filled with ethyl hexyl sebacate fluid in order to measure the capacitance at different temperatures.

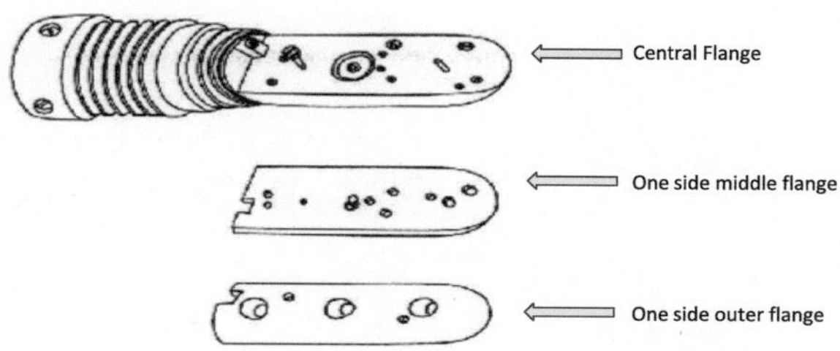

Fig. 2 Specimen holder for pressure and temperature measurement

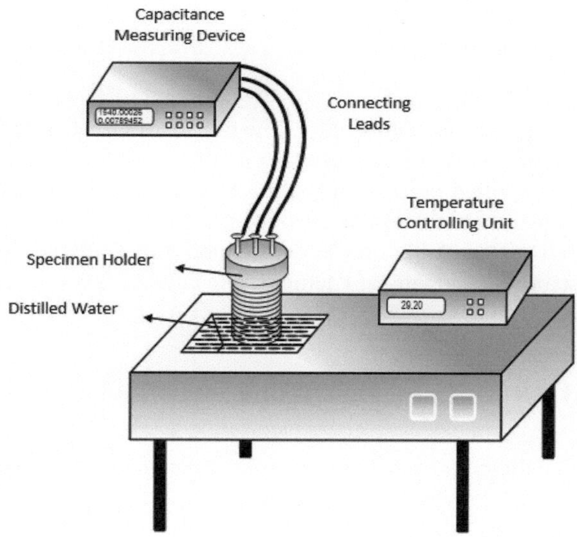

Fig. 3 Thermostatically controlled temperature bath

This vessel was submerged in temperature controlled distilled water. Each time the temperature was altered, as the specimens were in the middle of the holder, needed around 30 min to stabilize. In Fig. 3, an experimental setup is displayed.

2.2.2 Dielectric Measurement at Various Pressures

To investigate the dielectric properties under varying pressure conditions, the specimen holder was positioned within a high-pressure vessel constructed from maraging steel, designed to accommodate the holder. The pressure was generated utilizing a conventional high hydrostatic pressure system. A block diagram illustrating the pressure generation system is presented in Fig. 4. The pressure was transmitted through ethyl hexyl sebacate fluid, chosen for its non-reactivity with the specimen and its stable viscosity across the relevant pressure range. Surrounding the pressure vessel was a copper jacket, which included a copper tube connected to a temperature bath

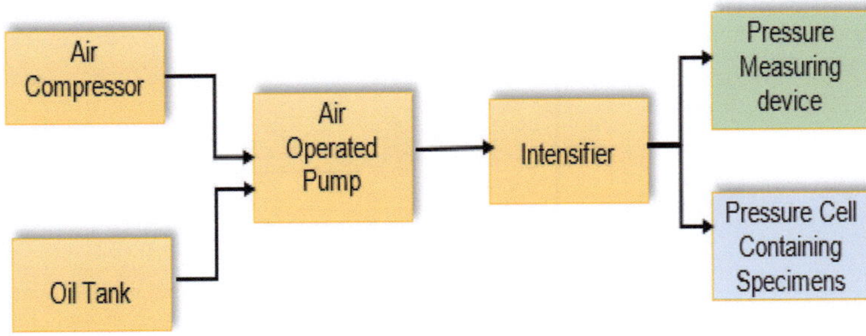

Fig. 4 Block diagram to measure pressure dependent dielectric properties

to regulate the temperature. The impact of pressure on the Chromel-Alumel thermocouple was minimal, resulting in a negligible temperature correction for this study. The applied pressure was measured using the national hydraulic primary pressure standard (NPLH1) [38], achieving a measurement uncertainty of less than 0.013%. A duration of 40 min was determined to be sufficient for the system to reach thermal stability following any temperature adjustments.

3 Result and Discussion

The dielectric constants (K) were calculated from capacitance (in pF).

3.1 Effect of Variation of Concentration of PT on Temperature and Pressure Dependent Dielectric Properties

Dielectric constants (K) of specimen prepared by varying the concentration of PT were measured at arbitrary temperature 283, 288, 293, 298, 303, 308, 313, 318, and 323 K. The plot between relative dielectric constant (K*T* /K*TO*) and temperature is shown in Fig. 5, where K*T* is dielectric constant at any temperature and K*TO* is at 293 Kelvin.

It is observed that slope of curve for PMN with 10% PT is positive, showing positive temperature coefficient of dielectric constant. And slopes of curves for 27 and 35% PT are nearly zero, showing very less temperature coefficient of dielectric constant.

The plot between relative dielectric constant (K*P*/K*PO*) and pressure is shown in Fig. 6, where K*P* is dielectric constant at any pressure and K*PO* is at atmospheric pressure.

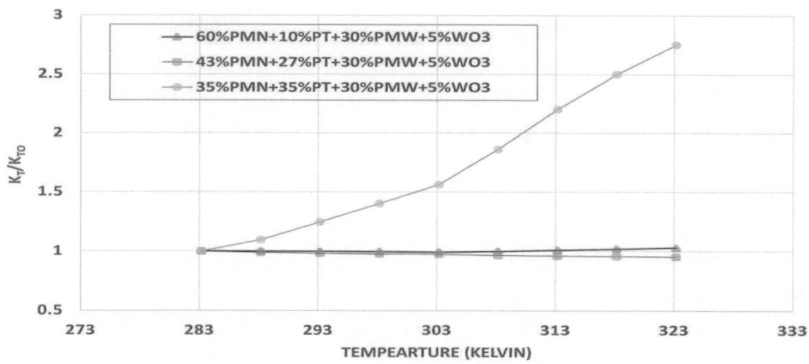

Fig. 5 Plot of variation of relative dielectric constant (K*T* /K*TO*) with temperature of (1-x–y)PMN-xPT-yPMW + zWO3 specimen)

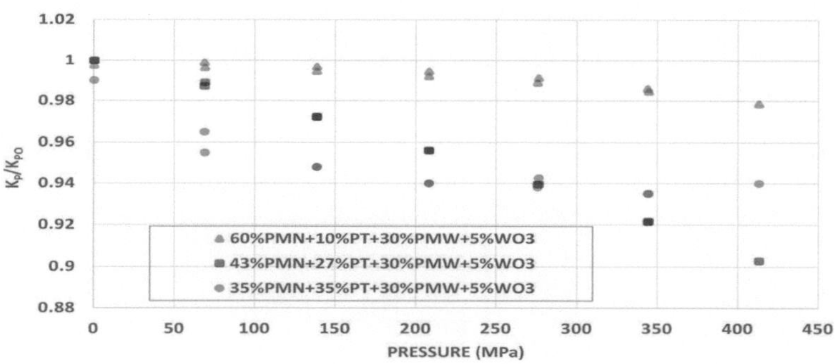

Fig. 6 Plot of variation of relative dielectric constant (K*P*/K*PO*) with Pressure of (1-x-y)PMN-xPT-yPMW + zWO3 specimen

Slopes of curves of containing 27 and 35% PT are negative but slope of 35% PT is lesser than that at 27%, showing less negative pressure coefficient of dielectric constant. And slope for specimen having 10% PT is nonlinear, showing negative nonlinear pressure coefficient of dielectric constant.

3.2 *Effect of Variation of Excess Concentration of WO3 on Temperature and Pressure Dependent Dielectric Properties*

Dielectric constants (K) of specimen prepared by varying the concentration of excess WO3 were calculated as a function of temperature. Plot between relative dielectric constant (K*T*/K*TO*) and temperature is shown in Fig. 7, where K*T* is dielectric constant at any temperature and K*T O* is at 293 K. It is observed that slope of all curves is nearly equal and almost zero, showing very less temperature coefficient of dielectric constant.

The plot between relative dielectric constant (K*P*/K*PO*) and pressure is shown in Fig. 8, where K*P* is dielectric constant at any pressure and K*PO* is at atmospheric pressure.

The Slopes of all curves are negative but slope of specimen having 5% WO3 is more than others, indicating a higher negative pressure coefficient of dielectric constant. It is evident that pressure and temperature coefficient of dielectric constant is optimized by addition of excess 5% of WO3 in 43%PMN-27%PT-30%PMW relaxor system. Because specimen 43%PMN-27%PT-30%PMW + 5%WO3 exhibits a moderate pressure coefficient and virtually zero temperature coefficient of dielectric constant, it appears to be acceptable for pressure gauging.

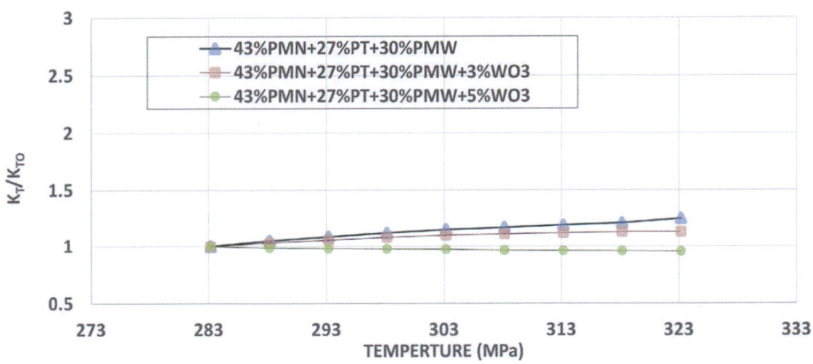

Fig. 7 Plot of variation of relative dielectric constant (K*T*/K*TO* with temperature of 43%PMN-27%PT-30%PMW + zWO3 specimen

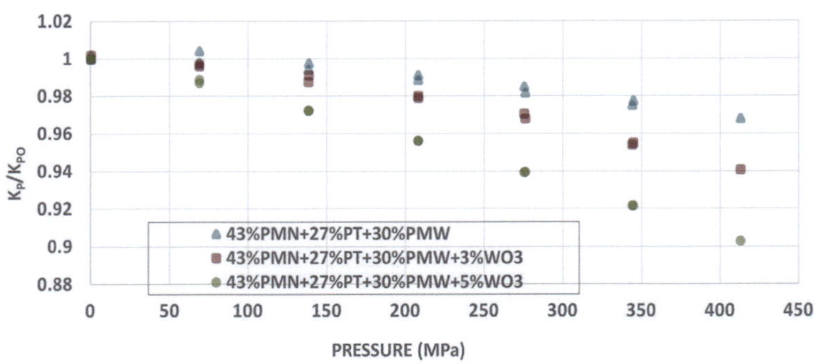

Fig. 8 Plot of variation of relative dielectric constant (K*P* /K*PO*) with pressure of 43%PMN-27%PT-30%PMW + zWO3 specimen

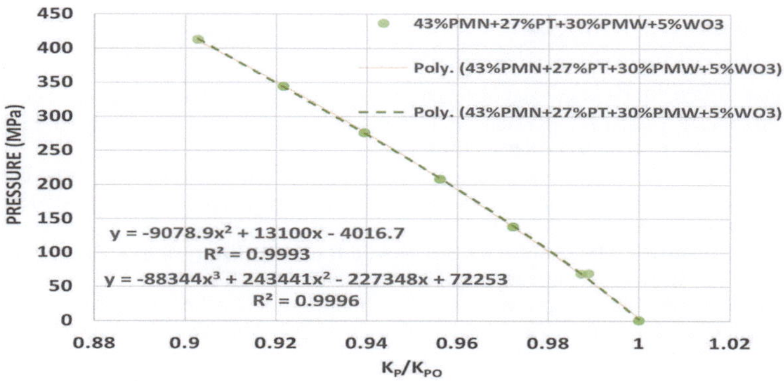

Fig. 9 Plot of variation pressure with relative dielectric constant (KP/KPO) of 43%PMN-27%PT-30%PMW + 5%WO3 specimen

3.3 Pressure Variation Trend Line

For pressure gauges, the specimen 43%PMN-27%PT-30%PMW + 5%WO3 is appropriate for pressure sensor. Figure 9 displays the formulae for fitting the specimen's pressure data to the quadratic and cubic trend lines.

When compared to a quadratic fitted equation, the cubic fitted equation's R2 value is greater and closer to 1. After 68 MPa of pressure, the trend line is almost linear.

4 Conclusion

The possibility of using PMN-PT-PMW relaxor as the sensing material in a pressure sensor is demonstrated by this work. The investigation comes to the conclusion that the dielectric permittivity of the PMN-PT-PMW relaxor and pressure have a quadratic/cubic connection. Among all the specimens analyzed, the 43%PMN-27%PT-30%PMW + 5%WO3 specimen is suitable to be used as a pressure sensing element to create a pressure measuring device, considering all the properties required for pressure measurement. Another advantage of this material is that it exhibits less hysteresis as the order of pressure changes. The results of this study would open the door for additional research on this material, with a focus on producing high pressure transducers up to 415 MPa and having less temperature coefficient and hysteresis.

Acknowledgements The author expresses sincere gratitude to Dr. Sanjay Yadav for his invaluable counsel and support.

References

1. Kumar V (2023) Pressure and its measurement. In: Aswal DK, Yadav S, Takatsuji T, Rachakonda P, Kumar H (eds) Handbook of metrology and applications. Springer, Singapore. https://doi.org/10.1007/978-981-99-2074-7_46
2. Kao KC (2004) Ferroelectrics, piezoelectrics, and pyroelectrics. In: Kao KC (eds) Dielectric phenomena in solids. Academic Press, ISBN 9780123965615, pp 213–282. https://doi.org/10.1016/B978-012396561-5/50014-1
3. Vijatovi´c Petrovi´c MM, Bobi´c JD, Stojanovi´c B (2008) History and challenges of barium titanate: part II. Sci Sinter 40. https://doi.org/10.2298/SOS0803235V
4. Bobic JD (2018) Magnetic, ferroelectric, and multiferroic metal oxides, review of the most common relaxor ferroelectrics and their applications 233–249. https://doi.org/10.1016/B978-0-12-811180-2.00011-6
5. Cheng ZY, Katiyar RS, Yao X (1997) Guo aquiang dielectric behavior of magnesium nibote relaxors. Phys Rev B 55(13):8165–8174. https://doi.org/10.1103/PhysRevB.55.8165
6. Kwata J, Uchino K, Nomura S (1979) Diffuse phase transition in lead zinc niobate. Ferroelectrics 22, 863–867. https://doi.org/10.1080/00150197908239443
7. Mizutani N, Lu CH, Shinozaki K, Kato M (1990) Formation of high temperature liquid phase during the sintering of PFW. J Am Ceram Soc (USA) 73(5):1214–1220
8. Hachiga T, Fujimoto S, Yasuda N (1987) The pressure and temperature de-pendent dielectric properties of PCN. J Phys D: Appl Phys 20:1291–1296
9. Bokov AA, Shonov VY (1990) Dielectric properties and diffusion of ferroelectric phase transition in lead indium niobate. Ferroelectrics 108:237–240. https://doi.org/10.1080/00150199008018763
10. Setter N, Cross L (1980) Pressure dependence dielectric properties of PST. Phys Status Solidi A, Appl Res 61:K71–K75. https://doi.org/10.1002/pssa.2210610157
11. Szwagierczak D, Kulawik J (2004) Thick film capacitors with relaxor dielectrics. J Eur Ceram Soc 24(7):1979–1985. https://doi.org/10.1016/S0955-2219(03)00358-3
12. Jiang Y, Niu X, Liang W, Jian X, Shi H, Li F, Zhang Y, Wang T, Gong W, Zhao X, Yao Y, Tao T, Liang B, Lu S (2022) Enhanced energy storage performance in Na0.5Bi0.5TiO3-based relaxor ferroelectric ceramics via compositional tailoring. Materials 15, 5881. https://doi.org/10.3390/ma15175881
13. Uchino K (1994) Relaxor ferroelectric devices. Ferroelectrics 151(1):321–330. https://doi.org/10.1080/00150199408244756
14. Hsu HS, Benjauthrit V, Zheng F, Chen R, Huang Y, Zhou Q, Shung KK (2012) PMN-PT–PZT composite films for high frequency ultrasonic transducer applications. Sens Actuators A179 121–124
15. Seoa I-T, Leeb T-G, Kima D-H, Hurb J, Kima J-H, Nahma S, Ryuc J, Choi B-Y (2016) Multi-layer piezoelectric haptic actuator with CuO-modified PZT-PZNN ceramics. Sens Actuators A 238:71–79
16. Damjanovic D, Muralt P, Setter N (2001) Ferroelectric sensors. IEEE Sens J 1(3). https://doi.org/10.1109/JSEN.2001.954832
17. Yasuda N, Ueda Y (1989) Dielectric properties of PbFe1/2Nb1/2O3 under pressure. Ferroelectrics 95(1):147–151. https://doi.org/10.1080/00150198908245193
18. Yasuda N, Fujimoto S, Tanaka K (1985) Dielectric properties of Pb(Fe2/3W1/3)O3 under pressure. J Phys D: Appl Phys 18:1909–1917. https://doi.org/10.1088/0022-3727/18/9/021
19. Kumar V (2021) Studies on optimization of preparation methods of lead iron tungstate relaxor to use as a pressure sensing material. MAPAN-J Metrol Soc India 36(2):305–311. https://doi.org/10.1007/s12647-021-00470-0
20. Kumar V, Yadav S (2017) Studies on the use of lead iron tungstate- ead titanate relaxor binary system as a pressure sensing material. Sens Actuators A: Phys 258:101–104. https://doi.org/10.1016/j.sna.2017.02.030
21. Kumar V, Kamlesh K. Jain, Kashyap SC (2000) Dielectric study of ferroelectric relaxor under high pressure; 43rd DAE-Solid State Physics Symposium 27–31, Bilaspur, MP

22. Kumar Y, Kumar V, Kamlesh K. Jain, Kashyap SC (1999) A capacitive pressure gauge as a reliable transfer pressure standard. Sens Actuators, B 55:217. https://doi.org/10.1016/S0925-4005(99)00088-X
23. Kumar V (2024) Studies on pressure and temperature dependent dielectric properties of lead magnesium niobate-lead titanate binary relaxor to use as a pressure sensing material. Recent Advances in Metrology. ADMET 2022. Lecture Notes in Mechanical Engineering. Springer, Singapore. https://doi.org/10.1007/978-981-99-4594-8_21
24. Pel´aiz-Barranco A, Calder´on-Pin˜ar F, Garc´ıa-Zald´ıvar O, Gonz´alez-Abreu Y (2012) Relaxor Behaviour in Ferroelectric Ceramics. In: Advances in Ferroelectrics. London, United Kingdom: Intech Open. https://www.intechopen.com/chapters/4127310.5772/52149
25. Bokov AA, Ye ZG (2012) Dielectric relaxation in relaxor ferroelectrics. J Adv Dielectr 2(2):1241010. https://doi.org/10.1142/S2010135X1241010X
26. Baek JG, Isobe T, Senna M (1997) Synthesis of pyrochlore free 0.9Pb (Mg1/3Nb2/3)O3-0.1PbTiO3 ceramics via a soft mechanochemical route. J Am Ceram Soc 80(4):973–981. https://doi.org/10.1111/j.1151-2916.1997.tb02929.x
27. Sato Y, Kanai H, Yamashita Y (1996) Effects of silver and palladium doping on the dielectric properties of 0.9Pb (Mg1/3Nb2/3)O3–0.1 PbTiO3 ceramic. J Am Ceram Soc 79(1):261–265. https://doi.org/10.1111/j.1151-2916.1996.tb07907.x
28. Yimnirun R, Ananta S, Laoratakul P (2004) Dielectric properties of ceramics in lead zirconate titanate-lead magnesium niobate system. Songk lanakarin J Sci Technol 26(4):529–536
29. Chen J, Fan H, Ke S, Chen X, Yang C, Fang P (2009) Relaxor behavior and dielectric properties of lead magnesium nio bate-lead titanate thick films prepared by electrophoresis deposition. J Alloy Compd 478:853–857. https://doi.org/10.1016/j.jallcom.2008.12.041
30. Politova ED, Ivanova VV, Kaleva GM, Mosunov AV, Prutchenko SG, Yu S, Stefanovich AG, Segalla, AL, Kholkin V
31. Shvartsman V (2004) Dielectric and piezoelectric properties of Pb(Mg 1/3Nb2/3)O3–PbTiO3–Pb(Mg1/2 W1/2)O3 ceramics. Inorg Mater 40(9):998–1005. https://doi.org/10.1023/b:inma.0000041336.13046.45
32. Pouya Moetakefa Z (2008) Synthesis of pyrochlore free PMN–PZT ceramics via a seeding method. Sens Actuators A 141:463–470. https://doi.org/10.1016/j.sna.2007.10.005
33. Jun SG, Kim NK (2000) Dielectric properties of PFW-PMN relaxor sys-tem prepared by B-site precursor method. J Mater Sci 35:2093–2097. https://doi.org/10.1023/a:1004711728961
34. Johnsson M, Lemmens P (2007) Crystallography and chemistry of perovskites. Handbook of magnetism and advanced magnetic materials. https://doi.org/10.1002/9780470022184.hmm411
35. hart M, Somayazulu M, Cohen RE, Ganesh P, Dera P, Mao HK, Hemley RJ, Ren Y, Liermann P, Wu Z (2008) Origin of morphotropic phase boundaries in ferroelectrics 451. https://doi.org/10.1038/nature06459
36. Isupov VA (2003) Nature of physical phenomena in ferroelectric. Phys Solid State 45(6):1107–1111. https://doi.org/10.1134/1.1583909
37. Abdulova AS, Kaleva GM, Prutchenko SG, Segalla AG, Politova ED (2003) Effect of nonstoichiometry on the structure and dielectric properties of PbMg1/3Nb2/3O3–PbTiO3 ceramics inorganic materi-als 39(4):398–403. https://doi.org/10.1023/A:1023288001933
38. Swartz SL, Shrout TR (1982) Fabrication of perovskite lead magnesium nio-bate. Mater Res Bull 17(10):1245–1250. https://doi.org/10.1016/0025-5408(82)90159-3
39. Yadav S, Om Prakash VK, Gupta AK (2007) Bandyopadhyay, the effect of pressure-transmitting fluids in the characterization of a controlled clearance piston gauge up to 1 GPa. Metrologia 44(3):222–233. https://doi.org/10.1088/0026-1394/44/3/009

Experimental Investigation and Multi-Variable Optimization of Electric Discharge Machining Process for AISI420 Stainless Steel

Ajay Pratap Singh, Harish Taluja, Sudhir Kumar, and Arpita Gupta

Abstract In today's competitive industrial environment, there is a demand for precise material manipulation. Electric discharge machining is an effective technology for working with difficult-to-mill materials. This research aims to investigate the relationship between machining parameters and process outcomes. To analyze the material removal rate, kerf width, and surface roughness of AISI420 stainless steel, we have selected four controllable machining parameters: the gap voltage, current, pulse on time, and pulse off time. The experiment will be conducted using the Taguchi technique orthogonal array (L16), and we will use the response surface approach to identify the best process parameters for reducing surface roughness (SR) and kerf width while maximizing material removal rate (MRR). We will use analysis of variance to assess the effectiveness of the mathematical models we are using. This research involves an extensive investigation to optimize responses simultaneously using the weighted grey relational approach. The findings showed that this methodology resulted in improved output responses during the EDM process, with a grey relational grade of 0.8226. The decision-making process for multi-objective optimization heavily depends on how weights are assigned to responses and the chosen optimization method. The experimental results demonstrate that the suggested strategy effectively enhances the overall performance of the machining process.

A. P. Singh (✉)
Ajay Kumar Garg Engineering College, Ghaziabad, Uttar Pradesh, India
e-mail: ajay.pratap.singh.86@gmail.com

H. Taluja
HRIT Group of Institutions, Ghaziabad, Uttar Pradesh, India
e-mail: harishtaluja@gmail.com

S. Kumar
Government Polytechnic Uttawar, Palwal, Haryana, India
e-mail: skmeiitdhn@gmail.com

A. Gupta
Dronacharya Group of Institutions, Greater Noida, Uttar Pradesh, India
e-mail: guptaarpita599@gmail.com

S. Yadav et al. (eds.), *Advances in Metrology*, Lecture Notes in Mechanical Engineering,
https://doi.org/10.1007/978-981-96-6418-4_12

Keywords Weighted grey relational analysis · Kerf width · Electrical discharge machining · Material removal rate · Analysis of variance · Surface roughness

1 Introduction

Advanced Machining Processes (AMP) goes beyond conventional methods to shape materials. A wedge-shaped tool is used, and chips are formed by generating contact stresses to remove excess material using mechanical, electrical, chemical, thermal, or combined energies [1]. These procedures' advantages improved exactness, handling materials challenging fir conventional approaches and shaping intricate configurations precisely. One such technique is electrical discharge machining (EDM), which has recently attracted significant attention due to its capacity to work with hard-to-cut materials and create complex geometries [2]. Chromium rich stainless steels (12–18%) deliver durable mechanical performance, applied in dental prosthetics, vehicle parts, pressure vessels, and industrial chemical blades [3]. Figure 1 elucidates EDM's material removal process. Ionized dielectric medium between electrodes, energizes by voltage, cause intense electrical discharge and localized heating, enabling phase transition (melting/evaporation) of the workpiece. Consequently, the procedure involves the removal of material from the workpiece [4]. Scholars globally have been dedicated to enhancing the process parameters of the EDM technique. Advancements in EDM have facilitated the production of intricate components, the adaptation of molds and dies, and micro-products. Rajmohan et al. performed an investigation on micro-EDM utilizing 2205 DSS. The Taguchi approach was employed to optimize the process parameters for various responses. Additionally, KW was further analyzed using the SEM [5]. Pramanik et al. investigated 2205 DSS process production utilizing W-EDM and utilized the RSM approach to plan and conduct experiments. During the trial, MRR showed an increase, whereas SR exhibited a decline [6]. Payla et al. investigated the W-EDM approach to reduce power consumption in their research, emphasizing the impact of several process parameters on the method's output and energy efficiency [7]. Hanif et al. researched the impact of dielectric fluid, electrode polarity, discharge current, and spark gap on MRR and SR during steel machining. They utilized RSM and GRA techniques. Their findings indicated that electrode polarity had the most significant effect [8]. Using OVAT, Gurpreet Singh employed assessed copper tool machining parameters for H13 steel, measuring efficacy via MRR, TWR, and SR, as critical output response [9]. Ahuja et al. explored input response dynamics in Mg alloy machining, employed RSM and the desirability function to evaluate multi-objective outcome comprehensively [10].

Le V.T. investigated few factors that impact the surface characteristics of SKD61 steel after it is machined using PMEDM and tungsten carbide powder, including peak current (Ip), pulse on time (Ton), and powder concentration (Cp). The surface parameters under consideration were SR, surface micro-hardness, and surface morphology [11]. Vinay et al. revealed DenseNet model flawless 100% accuracy in training/testing, whereas MNB model attained marginally reduced results (99.13% in training

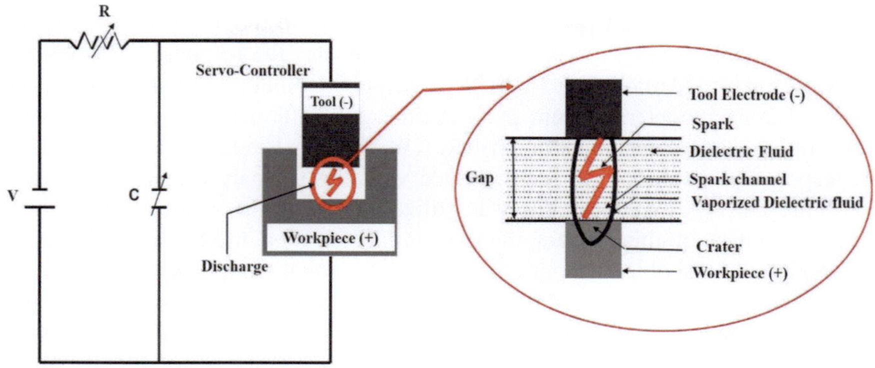

Fig. 1 Mechanism of material erosion in sinking-EDM [4]

and 98.98% in testing), emphasizing comparative model precision [12]. Karthik et al. leverage a Taguchi-based L27 OA with EDM factors for their investigation. They integrated entropy technique weights with CODAS approach to optimize the input parameters [13]. Cao et al. exploited micro-EDM to enhance the aspect ratio of fabricated micro-holes by optimizing the frequency of bubble production, resulting in more accurate and efficient machining [14]. Mehrvar and Mirak conducted a comparison of the surface of CMSX-4 nickel-based superalloy following the processes of grinding, WEDM, and ECM. ECM has demonstrated favorable outcomes, rendering it a viable substitute for producing components with sleek surfaces and reduced flaws [15]. Surface attributes and input factors under varying tool materials were analyzed by Adel et al. utilized ML models for surface condition predictions [16]. Shastri and Mohanty conducted an investigation in which they utilized different electrodes for EDM on Nimonic. The impacts were assessed by variance analysis and scanning electron microscopy [17]. Bhosale et al. assessed machinability of sintered PM ferrous parts using clay matrix EDM. They discovered that the content of clay, electric current, duration of pulses, and voltage had an impact on the erosion of the electrode, the pace at which material was removed, and the roughness of the surface. Prior to employing the combined GRA-TOPSIS framework, the system underwent optimization [18]. Dubey et al. explored AWJM parameter optimization [P, AFRE, SoD and TRS] using RSM to reduce PT, SR, and increasing hardness HRC [25]. Dubey and Kumar employed RSM-BBD (three parameters, three levels) and grey entropy weight (GEW) to analyze MRR, SR, and KTA, finding SoD most influential and TRS least [26].

Upon review of the existing literature and leveraging the author's expertise, it has come to our attention that there is a lack of documented optimization of EDM through the use of specific parameters to achieve desired responses by incorporating the weighted Grey Relational Analysis (w-GRA) method for AISI420 stainless steel. Consequently, this ongoing investigation seeks to comprehensively scrutinize and optimize the material removal rate (MRR), kerf width (KW), and surface roughness

(SR) concurrently via the w-GRA method. Adhering to the Taguchi L16 orthogonal array method, this experimental inquiry evaluates the responses utilizing the Response Surface Methodology for MRR, KW, and SR. Notably, this study centers on four key EDM parameters: current, pulse on time, pulse off time, and gap voltage, with the resulting outputs encompassing MRR, KW, and SR. The methodology included 3D response plots tailored to each targeted variables to analyze their spatial relationships. The ANOVA technique was utilized to assess the significance of these factors, and through this process, the most influential model terms were identified. Conventional multi-objective optimization assigns equal weightings, but industrial applications necessitate tailored prioritization. Utilizing w-GRA, this research optimizes parameters, verified via experimental validation. Results hold practical relevance for EDM scholars and industries. The novel integration of weighted analysis and empirical verification advances existing literature offering a scalable framework for precision engineering and adaptive manufacturing solutions.

2 Experimental Set-up and Measurements

The experimental setup utilized a sinking-EDM machine, and AISI420 stainless steel was chosen as the material for the experiment. This selection was made due to the material's superior properties to ferritic and austenitic stainless steels, making it well-suited for achieving the study's objectives. The workpiece used in the experiment was AISI420 stainless steel, with dimensions of 45 mm × 20 mm × 10 mm. To carry out the machining process, a rectangular copper electrode tool measuring 20 mm × 12 mm × 40 mm was employed, chosen for its excellent thermal conductivity. Detailed information on the properties, chemical composition, and physical characteristics of the electrode material and AISI420 stainless steel can be found in Tables 1 and 2, respectively.

The cutting depth was consistently set at 0.50 mm. Critical parameters such as Current (Ip/ampere), Pulse on Time (Ton/μ-sec), Gap Voltage (A/volt), and Pulse on Time (Toff/μ-sec) were carefully selected. Measured metrics such as MRR, KW, and SR are essential for evaluation. Machining conditions variables (MCVs) were specifically chosen for investigation. The experimental design was generated using DESIGN EXPERT 13 software. After each experiment, a thorough cleaning of the workpiece and electrode was conducted. The MRR was calculated using Eq. 1. The experimental results are presented in Table 3 (Table 4).

Table 1 Copper (electrode) properties [19]

Electrode material	Density (gm/cc)	Specific heat (J/kg.K)	Thermal conductivity (W/mK)	Electrode resistivity (μ-ohms)	Hardness (BHN)
Copper	8.9	386	399	1.69	48

Table 2 Chemical composition and physical properties of AISI420 stainless steel [19]

Chemical composition of AISI420 stainless steel					
Chromium	Manganese	Silicon	Carbon	Sulphur	Phosphorous
31.208	0.310	0.384	0.23	0.030	0.040
Physical properties of AISI420 stainless steel					
Density (gm/cc)	Specific heat (KJ/kg.K)	Thermal conductivity (W/mK)	Hardness (HB)	Tensile strength (MPa)	Elongation %
7750	0.46	24.9	201–255	1120	18.2

Table 3 Machining parameters and their levels for design of experiments

Input parameters	Unit	Code	Levels with selected values				Experimental design
			1	2	3	4	
Gap voltage	Volt (V)	A	15	45	75	105	The investigations were carried out employing a Taguchi orthogonal array (L16 OA). The ideal process parameters that minimize KW and SR and maximize MRR were determined using RSM
Current	Ampere (A)	I_P	2	4	6	8	
Pulse on time	μsec	T_{on}	100	110	120	130	
Pulse off time	μsec	T_{off}	30	40	50	60	

$$MRR\,(g/\min.) = \frac{\text{Workpiece weight}_{\text{before machining}} - \text{Workpiece weight}_{\text{after machining}}}{\text{Time}_{\text{machining}}} \tag{1}$$

3 Methodology

The Design of Experiments (DOE) is a rigorous and systematic approach used in statistical research and engineering to uncover the factors impacting the behavior of a system or process. It involves thorough planning, execution, and analysis of controlled experiments [20]. In industrial contexts, efficiently utilizing machining facilities is essential to boost productivity and product quality. The Taguchi method is a powerful tool for optimizing process parameters, including those for machining operations [21]. Moreover, Response Surface Methodology (RSM) provides a robust statistical and mathematical framework for scientists, engineers, and technologists to improve processes, enhance product quality, and optimize system performance [22].

Table 4 Experimental data for MCV and MMP

Exp. No.	Machine Control Variables (MCV)				Response Parameter (avg.)		
	A	I_P	T_{on}	T_{off}	SR (μm)	KW (microns)	MRR (g/min)
1	15	2	100	30	2.42	312.33	2.85
2	15	4	110	40	2.47	314.86	2.81
3	15	6	120	50	2.52	317.39	3.02
4	15	8	130	60	2.57	319.93	3.23
5	45	2	110	50	2.63	326.97	4.02
6	45	4	100	60	2.42	327.63	3.75
7	45	6	130	30	2.73	334.01	3.97
8	45	8	120	40	2.52	334.67	3.7
9	75	2	120	60	2.84	342.1	5.31
10	75	4	130	50	2.88	345.62	5.29
11	75	6	100	40	2.42	345.39	4.31
12	75	8	110	30	2.47	348.91	4.28
13	105	2	130	40	3.04	358.72	6.26
14	105	4	120	30	2.84	360.36	5.76
15	105	6	110	60	2.63	360.03	5.72
16	105	8	100	50	2.42	361.68	5.22

RSM seamlessly integrates experimental data through the effective combination of DOE, regression modeling, and optimization techniques, bridging the gap between theory and practice [23].

$$S = f\,(X, Y, Z) \pm e_r \tag{2}$$

$$S = n_o + \sum\nolimits_{i=1}^{3} n_i B_i + \sum\nolimits_{i=1}^{3} n_{ii} B_i^2 + \sum\nolimits_{i=1}^{3} n_{ij} B_i B_j \tag{3}$$

In research, "Grey" details refer to information that is insufficient, questionable, or incomplete. To ensure accuracy and acceptability, it is crucial to collect extensive data that meets specific and measurable conditions. Traditional statistical techniques are employed for this purpose. Grey theory provides a satisfactory explanation for dealing with ambiguous and multi-input information. Grey Relational Analysis (GRA) seeks to identify the connections between a key aspect and all other relevant factors within a given system [24]. The MRR, SR, and KW have been optimized using GRA. The experiment aims to minimize the KW and SR, and MRR is to be maximized. The first step after that is pre-processing data, which entails changing the initial order to one that is similar. This is necessary because the output parameter test result may differ from other information flows in terms of variety, unit, and objective path.· As a result, the data associated with a set of progressions must be pre-processed. The information order for KW, SR, and MRR are pre-processed by the Eqs. 4 and 5.

$$K\prime_{ij} = \frac{L\prime_{ij} - \min_j L\prime_{ij}}{\max_i L\prime_{ij} - \min_j L\prime_{ij}} \tag{4}$$

$$K\prime_{ij} = \frac{\max_j K\prime_{ij} - L\prime_{ij}}{\max_i L\prime_{ij} - \min_j L\prime_{ij}} \tag{5}$$

Equation 6 outlines the computational framework for the GRC, which is applied in this study.

$$\xi\prime_{ij} = \frac{\min_i \min_j \left|K_i^o - K\prime_{ij}\right| + \zeta \max_i \max_j \left|K_i^o - K\prime_{ij}\right|}{\left|K_i^o - K\prime_{ij}\right| + \zeta \max_i \max_j \left|K_i^o - K\prime_{ij}\right|} \tag{6}$$

The GRG is derived by averaging the GRC, and this metric is subsequently applied in Eq. 7 to assess MMP. A bigger GRG shows remarkable multi-response.

$$Q_{i,j} = \max\{O_{i,j,1}, O_{i,j,2}, \ldots\ldots\ldots O_{i,j,k}\} - \min\{O_{i,j,1}, O_{i,j,2}, \ldots\ldots\ldots O_{i,j,k}\} \tag{7}$$

$$\omega_i = \frac{\sum_{j=1}^{p} Q_{i,j}}{\sum_{i=1}^{m}\sum_{j=1}^{p} Q_{i,j}} \tag{8}$$

4 Results and Discussion

In Fig. 2, it is illustrated that a longer pulse on time (T_{on}) results in more powerful electrical discharges, leading to the creation of larger and deeper craters on the surface of the workpiece and a higher SR. Conversely, a shorter Ton generates smaller and shallower craters, thereby enhancing the surface finish. Additionally, a higher current (Ip) amplifies the electrical discharges, causing the formation of more extensive and deeper craters, as well as a higher SR. A higher Ip also increases the material removal rate (MRR) and results in larger craters, indicating a wider kerf width (KW). The gap voltage (A) has a significant impact on the lateral vibration of the wire and the breakdown distance in EDM, affecting the KW. Narrowing the gap voltage tends to reduce the KW. Moreover, increasing the pulse on time (T_{on}) generally leads to a decrease in the MRR.

With the help of Eq. (6), GRC are derived, and the response weight are established via the developed framework. Comparable procedures are utilized to quantify the GRC_{mean} for the remaining parameters at every hierarchical level. The GRC ranges

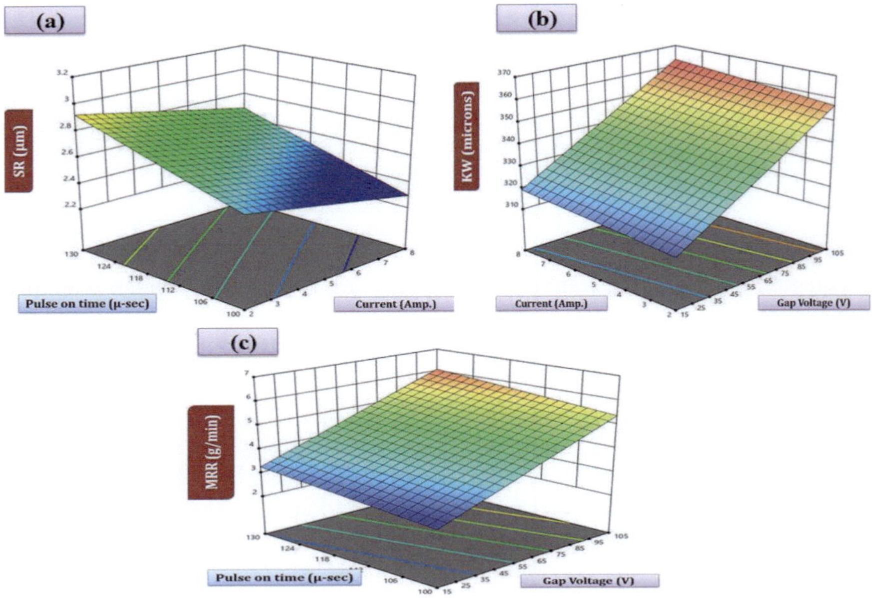

Fig. 2 Three-dimensional model plot for various MMP's vs MCV's

widen proportionality with heightened parametric influence on a response. However, negligible response shifts during MCV changes suggest an absence of MCV-MMP interdependence in the defined range, justifying a weight factor of zero (Table 5).

Given the sensitivity of responses to MCV variations, prioritizing higher weighting coefficients is advisable. Within the sixteen experimental trials, experiment 1 yielded the highest GRG (0.8226) approaching unity. The computed weighted GRG values derived using the specified methodology are tabulated in Table 6.

Table 7 outlines the average GRG values (GRG_{mean}) corresponding to distinct MCV levels. The magnitude of influence for parameters A, I_p, T_{on}, and T_{off} in EDM is quantified by the GRG_{max}-GRG_{min} range, where T_{on} registers the highest value (0.1996), as illustrated in the table. In EDM, the hierarchy of influence is ranked as 0.1996 > 0.1075 > 0.0341 > 0.0311, with T_{on} exhibiting the highest impact. The GRG_{mean} graph (Fig. 3) elaborates on the MCV relationship, with the dashed line signifies the mean GRG value for the comprehensive analysis. A strong positive correlation exits between improved multi-response performance and increased GRG values. The configuration A_1 I_{p4} $T_{on1}T_{off2}$ achieves the maximum GRG. A statistical assessment via ANOVA was performed to evaluate the critical MCVs influencing the MMP, highlighting T_{on} as the most significant variable. The prioritization of factors by ANOVA significance is T_{on} > A > T_{off} > I_p.

$$\hat{a} = a_m + \sum_{i=1}^{p} \bar{a} - a_m \tag{9}$$

Table 5 Normalized values and deviation sequence of responses

Exp. No.	Normalized values			Deviation sequence		
	SR	EWR	MRR	SR	EWR	MRR
1	1.0000	1.0000	0.0000	0.0000	0.0000	1.0000
2	0.9194	0.9487	0.0548	0.0806	0.0513	0.9452
3	0.8387	0.8975	0.1123	0.1613	0.1025	0.8877
4	0.7581	0.8460	0.1699	0.2419	0.1540	0.8301
5	0.6613	0.7033	0.3863	0.3387	0.2967	0.6137
6	1.0000	0.6900	0.3123	0.0000	0.3100	0.6877
7	0.5000	0.5607	0.3726	0.5000	0.4393	0.6274
8	0.8387	0.5473	0.2986	0.1613	0.4527	0.7014
9	0.3226	0.3968	0.7397	0.6774	0.6032	0.2603
10	0.2581	0.3254	0.7342	0.7419	0.6746	0.2658
11	1.0000	0.3301	0.4658	0.0000	0.6699	0.5342
12	0.9194	0.2588	0.4575	0.0806	0.7412	0.5425
13	0.0000	0.0600	1.0000	1.0000	0.9400	0.0000
14	0.3226	0.0267	0.8630	0.6774	0.9733	0.1370
15	0.6613	0.0334	0.8521	0.3387	0.9666	0.1479
16	1.0000	0.0000	0.7151	0.0000	1.0000	0.2849

Equation 9 derives the $GRG_{estimated\ optimal}$, with $GRG_{total\ mean}$ denoting the average GRG value at the optimal configuration, and p indicating the total number of MCVs influencing the observed machining outcomes. Experimental trails are carried out and ensuing the data, outlined in Table 8, highlights the gains achieved in various performance indicators using the w-GRA framework. The experimental evidence corroborates the efficacy of the suggested technique in elevating machining outcomes.

5 Conclusions

The research establishes that MCV achieved Optimal performance in sinking EDM, maximizing MRR, and reducing KW and SR. Interdependencies among the MMP's metrics (SR, KW, and MRR) are investigated via w-GRA framework. The core conclusions are outlined as follows:

1. T_{on} was shown the biggest impact on machining performance, followed by A, T_{off}, and I_p.

Table 6 GRC, Mean, and Weighted method (GRG values and rank)

Exp. No	GRC			Mean method		Weighted method	
	SR	EWR	MRR				
				GRG	Rank	GRG	Rank
1	1.0000	1.0000	0.3333	0.7778	1	0.8226	1
2	0.8611	0.9070	0.3460	0.7047	2	0.7411	4
3	0.7561	0.8298	0.3603	0.6487	5	0.6794	6
4	0.6739	0.7645	0.3759	0.6048	7	0.6314	8
5	0.5962	0.6276	0.4490	0.5576	12	0.5871	12
6	1.0000	0.6173	0.4210	0.6794	3	0.7542	3
7	0.5000	0.5323	0.4435	0.4919	16	0.5177	16
8	0.7561	0.5248	0.4162	0.5657	10	0.6200	10
9	0.4247	0.4532	0.6577	0.5118	14	0.5394	14
10	0.4026	0.4257	0.6530	0.4937	15	0.5208	15
11	1.0000	0.4274	0.4834	0.6369	6	0.7270	5
12	0.8611	0.4028	0.4796	0.5812	8	0.6575	7
13	0.3333	0.3472	1.0000	0.5602	11	0.5924	11
14	0.4247	0.3394	0.7849	0.5163	13	0.5549	13
15	0.5962	0.3409	0.7717	0.5696	9	0.6269	9
16	1.0000	0.3333	0.6370	0.6568	4	0.7569	2

Table 7 GRG_{mean} (weighted method) by factor level

	1	2	3	4	Δ (Max–Min)	Rank
A	0.7186	0.6198	0.6112	0.6328	0.1075	2
I_p	0.6354	0.6427	0.6378	0.6664	0.0311	4
T_{on}	0.7652	0.6531	0.5984	0.5656	0.1996	1
T_{off}	0.6382	0.6701	0.6361	0.6380	0.0341	3

Mean value of GRG (weighted method): 0.6456
Note *Optimal level of parameter

2. According to findings obtained from the GRA (Mean approach) and GRA (Weighted technique) investigations, the factor with the greatest impact was determined to be T_{on} (pulse current), with $GRG_{experiment}$ values of 0.7778 and 0.8226, respectively.
3. ANOVA revealed that factor T_{on} had the most significant impact on the material machining performance and the ANOVA results revealed the following hierarchy of statistical impact: $T_{on} > A > T_{off} > I_p$
4. Empirical outcomes revealed a significant improvement in the process's grey relational grade through the implementation of the developed frameword. The growth in grey relational grade was measured at 0.1623.

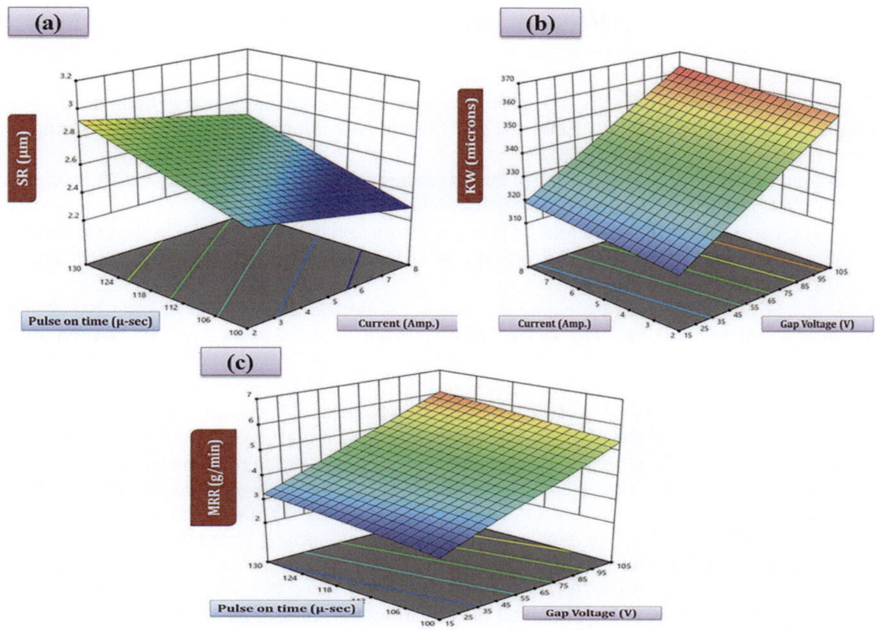

Fig. 3 Response graph of EDM parameters on GRG (weighted method)

Table 8 Results of machining performance using initial and optimal machining parameter

MCV	Initial machining parameters	Optimal machining parameters	
		Predicted	Experimental
	$A_4\ I_{p1}\ T_{on4}T_{off2}$	$A_1\ I_{p1}\ T_{on1}T_{off2}$	$A_1\ I_{p1}\ T_{on1}T_{off1}$
SR (μm)	2.88	–	2.42
KW	319.62		312.33
MRR (g/min)	2.63		2.85
GRG	**0.6603**	**0.8083**	**0.8226**
$GRG_{improvemnet}$	**0.1623**		

References

1. Okokpujie IP, Tartibu LK (2023) Overview of advanced machining process. In: Modern optimization techniques for advanced machining. Studies in systems, decision and control, vol 485, Springer, Cham. https://doi.org/10.1007/978-3-031-35455-7_1
2. Gohil V, Puri YM (2016) Turning by electrical discharge machining: a review. SAGE Publishing, vol. 231(2):195–208
3. Kumar S, Ghoshal SK, Arora PK et al (2024) Unlocking AISI420 martensitic stainless steel's potential: precision enhancement via S-EDM with copper electrodes and multivariate optimization. Arab J Sci Eng. https://doi.org/10.1007/s13369-024-08711-5

4. Kumar S et al (2021) Multi-variable optimization in die-sinking EDM process of AISI420 stainless steel. Mater Manuf Processes 36:572–582. https://doi.org/10.1080/10426914.2020.1843678
5. Rajmohan K, Kumar AS (2017) Experimental investigation and prediction of optimum process parameters of micro-wire-cut EDM of 2205 DSS. Int J Adv Manuf Technol 93:187–204. https://doi.org/10.1007/s00170-016-8615-3
6. Pramanik A et al (2018) Processing of duplex stainless steel by WEDM. Mater Manuf Processes 33:1559–1567. https://doi.org/10.1080/10426914.2018.1453165
7. Payla A et al (2019) Investigations on power consumption in WEDM of EN31 steel for sustainable production. Mater Manuf Processes 34:1855–1865. https://doi.org/10.1080/10426914.2019.1683577
8. Hanif M et al (2019) Investigating the effects of electric discharge machining parameters on material removal rate and surface roughness on AISI D2 steel using RSM-GRA integrated approach. Int J Adv Manuf Technol 101:1255–1265. https://doi.org/10.1007/s00170-018-3019-1
9. Singh G et al (2019) Analysis of electrical discharge machining parameters for H13 steel using OVAT technique. Eng Res Express 1:025031. https://doi.org/10.1088/2631-8695/ab52dc
10. Ahuja N et al (2020) Experimental investigation and optimization of wire electrical discharge machining for surface characteristics and corrosion rate of biodegradable Mg alloy. J Mater Eng Perform 29:4117–4129. https://doi.org/10.1007/s11665-020-04905-8
11. Le VT (2021) Influence of processing parameters on surface properties of SKD61 steel processed by powder mixed electrical discharge machining. J Mater Eng Perform 30:3003–3023. https://doi.org/10.1007/s11665-021-05584-9
12. Vakharia V et al (2022) Experimental investigations and prediction of WEDMed surface of nitinol SMA using SinGAN and DenseNet deep learning model. J Market Res 18:325–337. https://doi.org/10.1016/j.jmrt.2022.02.093
13. Karthik Pandiyan G, Prabaharan T, Jafrey Daniel James D, et al (2022) Machinability analysis and optimization of electrical discharge machining in AA6061-T6/15wt.% SiC composite by the multi-criteria decision-making approach. J Mater Eng Perform 31:3741–3752. https://doi.org/10.1007/s11665-021-06511-8
14. Cao P, Tong H, Li Y et al (2023) Interelectrode gas–liquid-solid three-phase flow analysis and simulation for drilling holes with high aspect ratio by micro-EDM. Int J Adv Manuf Technol 128:5261–5276. https://doi.org/10.1007/s00170-023-12220-4
15. Mehrvar A, Mirak A (2023) Comparative evaluation of surface integrity of CMSX-4 nickel-based superalloy after grinding, wire electrical discharge machining, and electrochemical machining processes. J Mater Eng Perform. https://doi.org/10.1007/s11665-023-08194-9
16. Abbas AT et al (2023) A machine learning perspective to the investigation of surface integrity of Al/SiC/Gr composite on EDM. J Manuf Mater Process 7(5):163. https://doi.org/10.3390/jmmp7050163
17. Shastri RK, Mohanty CP (2021) Sustainable electrical discharge machining of Nimonic C263 superalloy. Arab J Sci Eng 46:7273–7293. https://doi.org/10.1007/s13369-020-05211-0
18. Bhosale SB, Bhowmik S, Ray A (2022) Evaluation of machinability and recast layer analysis of ferrous clay composite through electric discharge machining process. Arab J Sci Eng 47:8523–8533. https://doi.org/10.1007/s13369-021-06337-5
19. Kumar S et al (2020) Optimization of process variables in electric discharge machining (EDM) using Taguchi methodology. Indian J Eng Mater Sci 27:819–825
20. Montgomary DC (2001) Design and analysis of experiments, 5th edn. Wiley, New York, ISBN: 0471316490, p.13–17.
21. Ross PJ (1996) Taguchi techniques for quality engineering: loss function, orthogonal experiments, parameter and tolerance design, 2nd edn. McGraw-Hill, New York, ISBN: 978-0070539587, pp 56–57
22. Box GEP, Wilson KB (1951) On the experimental attainment of optimum conditions. J R Stat Soc 13(1):1–45

23. Myers R, Montgomery D (1995) Response Surface Methodology. Process and Product Optimization Using Designed Experiments. Wiley, New York
24. Lin ZC, Ho CY (2003) Analysis and application of grey relation and ANOVA in chemical-mechanical polishing process parameters. Int J Adv Manuf Technol 21(1):10–14. https://doi.org/10.1007/s0017
25. Dubey AK, Kumar Y, Kumar S et al (2024) Parametric optimization of AWJM Using RSM-Grey-TLBO-based MCDM approach for titanium grade 5 alloy. Arab J Sci Eng. https://doi.org/10.1007/s13369-024-09500-w
26. Dubey AK, Kumar Y, Kumar S (2024) Optimizing parameters of AWJM for Ti-6Al-4 V grade 5 alloy using grey entropy weight method: a multivariable approach. J Inst Eng India Ser C. https://doi.org/10.1007/s40032-024-01139-8

New Metrological Challenges in a Sustainable Future

Sanjay Yadav

Abstract The pursuit of a sustainable future presents significant challenges and opportunities for metrology, the science of measurement. Accurate and reliable measurements are crucial for monitoring environmental changes, optimizing green technologies, and implementing sustainable practices. Metrological measurements are crucial in various sectors across numerous fields, say manufacturing, healthcare, oil and exploration, power (thermal, hydro, Nuclear), and defense to environmental monitoring and climate change. As the global community increasingly focuses on sustainability and this year UNESCO World Metrology Day (May 20, 20,024) was also celebrated as Metrology for Sustainability as a theme "We Measure Today for a Sustainable Tomorrow". Therefore, metrology faces new challenges and opportunities. Accurate and reliable measurements are essential for monitoring environmental changes, ensuring the quality and efficiency of green technologies, and implementing sustainable practices. This paper briefly highlights the emerging metrological challenges in achieving a sustainable future, emphasizing the need for advancements in measurement science to address environmental, technological, and societal demands. Furthermore, the paper emphasizes the importance of international collaboration and standardization in metrology. Organizations such as the International Bureau of Weights and Measures (BIPM) and the International Organization for Standardization (ISO) are instrumental in developing universally accepted measurement standards, facilitating data comparability, and enabling coordinated global efforts. Effective collaboration mechanisms, including joint research projects, interlaboratory comparisons, and capacity-building initiatives, are critical for advancing metrological practices and addressing disparities in technological capabilities and resources among countries. The integration of advanced technologies like artificial intelligence and the Internet of Things (IoT) into metrological practices is identified as a promising avenue for enhancing data collection and analysis. Overall, the paper underscores that international collaboration and standardization in metrology are indispensable for achieving sustainability goals, providing the foundation for informed decision-making and innovative solutions to global challenges.

S. Yadav (✉)
Formerly With CSIR—National Physical Laboratory, New Delhi, India
e-mail: ysanjay62@gmail.com

S. Yadav et al. (eds.), *Advances in Metrology*, Lecture Notes in Mechanical Engineering,
https://doi.org/10.1007/978-981-96-6418-4_13

Keywords Metrology · Sustainable developments · Metrological challenges · Advanced metrology

1 Introduction

The pursuit of a sustainable future is one of the most pressing challenges of our time. As the global community grapples with the consequences of climate change, resource depletion, and environmental degradation, the need for accurate and reliable data has never been more critical. Metrology, the science of measurement, is fundamental to this endeavor. It provides the precise measurements necessary for monitoring environmental changes, developing and deploying green technologies, and implementing sustainable practices across various sectors [1]. In an era defined by unprecedented technological advancements and complex environmental challenges, metrology faces new demands and opportunities. Accurate measurements are essential for understanding and mitigating the impacts of climate change, optimizing the efficiency of renewable energy systems, and ensuring the sustainability of manufacturing processes [2–6]. Moreover, as society increasingly adopts the principles of the circular economy, reliable measurements are crucial for assessing the quality and lifespan of recycled materials, thereby reducing waste and promoting resource reuse [7, 8].

The role of metrology extends beyond environmental monitoring and technological optimization [9, 10]. It is also vital for safeguarding public health and safety by ensuring the quality of air, water, and food, as well as assessing the safety of new materials and chemicals used in sustainable technologies [11–14]. Furthermore, the establishment of international measurement standards facilitates global collaboration, enabling countries to share data and best practices in addressing common sustainability challenges. Thus, the field of metrology must evolve to meet the demands of a sustainable future. This involves developing advanced measurement techniques, creating robust calibration standards, and fostering international cooperation. As the foundation of informed decision-making and innovative solutions, metrology will play an increasingly pivotal role in guiding the global community toward a more sustainable and resilient future [15–18]. This paper discusses the new metrological challenges in achieving sustainability (as shown in Fig. 1), highlighting the critical areas where advancements in measurement science are essential.

2 Environmental Monitoring and Climate Change

Environmental monitoring is a cornerstone in the fight against climate change. Precise measurement of various environmental parameters—such as atmospheric gases, temperature, sea levels, and ice cover—is essential for understanding and

Fig. 1 Metrological challenges in achieving sustainability

mitigating the impacts of global warming [11, 12]. This involves a complex interplay of advanced metrological techniques and robust calibration standards to ensure data accuracy and reliability over time. As mentioned, one of the most significant challenges in metrology for a sustainable future is the precise measurement of environmental parameters. Climate change, driven by greenhouse gas emissions, requires accurate monitoring of atmospheric gases, temperature, sea levels, and other critical indicators. Traditional measurement techniques often lack the precision and sensitivity needed to detect subtle changes over time. Advanced metrological methods, such as satellite-based remote sensing and in-situ sensor networks, are essential for providing high-resolution data to model and predict climate behavior. Some of the important issues in environmental monitoring and climate change are shown in Fig. 2 and are discussed thereafter.

2.1 *Atmospheric Monitoring*

The accurate measurement of greenhouse gases (GHGs) such as carbon dioxide (CO_2), methane (CH_4), and nitrous oxide (N_2O) is critical for tracking emissions and assessing the effectiveness of mitigation strategies [19]. Advanced spectroscopic techniques, such as satellite-based remote sensing and ground-based sensor networks, provide high-resolution data on GHG concentrations. Satellite instruments like NASA's Orbiting Carbon Observatory-2 (OCO-2) offer global coverage and enable the detection of even minor changes in atmospheric CO_2 levels. Ground-based networks, like the Integrated Carbon Observation System (ICOS) in Europe, complement these observations with localized, high-precision measurements.

Fig. 2 Some of the important issues in atmospheric monitoring and climate changes

2.2 Temperature and Sea Level

Monitoring global temperature trends is vital for detecting climate change. High-precision thermometers and satellite-based radiometers measure surface and atmospheric temperatures with increasing accuracy. Long-term datasets, such as those provided by NASA's Goddard Institute for Space Studies (GISS) and the National Oceanic and Atmospheric Administration (NOAA), are indispensable for identifying warming trends and informing climate models [20]. Sea level rise, driven by thermal expansion and melting ice caps, poses a significant threat to coastal communities. Satellite altimetry missions, like those conducted by the Jason series of satellites, measure sea surface height with centimeter accuracy. These measurements are crucial for predicting future sea level rise and planning coastal defense strategies.

2.3 Ice and Snow Cover

The measurement of ice and snow cover, particularly in polar regions, is essential for understanding the impacts of climate change on these sensitive areas. Satellite missions, such as the European Space Agency's CryoSat and NASA's ICESat, provide detailed data on ice thickness and volume. These measurements help scientists monitor the health of polar ice caps and glaciers, which are key indicators of global climate change [21].

2.4 Calibration and Standardization

Ensuring the accuracy of environmental measurements requires robust calibration standards. Instruments must maintain precision over long periods and under varying conditions. This necessitates the development of standardized calibration methodologies and the establishment of international reference standards. Organizations like the International Bureau of Weights and Measures (BIPM) and the World Meteorological Organization (WMO) play crucial roles in harmonizing these standards globally [22, 23]. Additionally, the calibration of these measurement instruments poses a significant challenge. Instruments must maintain accuracy over long periods and under varying environmental conditions. This requires the development of robust calibration standards and methodologies that can be applied globally. As climate change impacts intensify, the need for harmonized and universally accepted measurement standards becomes more critical.

3 Green Technologies and Renewable Energy

The transition to green technologies and renewable energy sources, such as solar, wind, and bioenergy, is essential for achieving a sustainable future. This shift demands precise and reliable measurements to optimize efficiency, ensure reliability, and minimize environmental impacts. Accurate measurement of energy production, storage, and consumption is vital for optimizing the performance of renewable energy systems [24]. Green technologies and renewable energy are depicted in Fig. 3 and discussed thereafter.

3.1 Solar Energy

Solar energy harnesses sunlight to generate electricity using photovoltaic (PV) cells. Accurate measurement of the efficiency of PV cells is critical for improving their design and maximizing energy output. This requires precise calibration of sensors that measure solar irradiance, temperature, and electrical output. Advances in metrology are needed to develop more sensitive and stable calibration techniques, ensuring consistent and accurate performance over the lifespan of solar panels. Furthermore, the spectral distribution of sunlight must be measured accurately to understand its impact on different types of PV materials and technologies [25, 26].

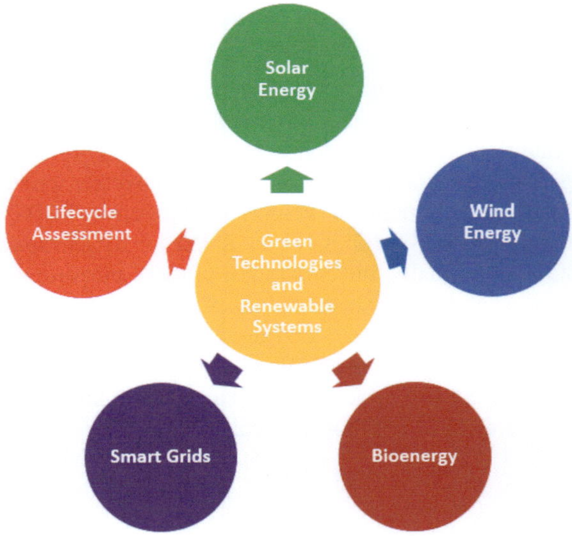

Fig. 3 Green technologies and renewable systems

3.2 *Wind Energy*

Wind energy relies on the conversion of kinetic energy from wind into mechanical energy using wind turbines. The efficiency and reliability of wind energy systems depend on accurate measurements of wind speed, direction, and turbulence [27, 28]. These parameters are influenced by local topography and atmospheric conditions, necessitating sophisticated metrological techniques. Lidar (Light Detection and Ranging) technology, for example, provides high-resolution wind measurements, enabling optimal placement and operation of wind turbines. Additionally, real-time monitoring and control systems for wind farms require continuous and precise data on wind conditions to maximize energy capture and minimize wear and tear on turbines [27–29].

3.3 *Bioenergy*

Bioenergy involves converting organic materials into energy through processes such as combustion, anaerobic digestion, and gasification. Accurate measurement of the calorific value, moisture content, and chemical composition of biomass is essential for optimizing these processes. Advanced analytical techniques, including spectroscopy and chromatography, are employed to characterize biomass feedstocks and monitor conversion processes. Metrology ensures that bioenergy systems operate efficiently and produce energy with minimal environmental impact [30, 31].

3.4 Smart Grids

The integration of renewable energy sources into the power grid presents additional metrological challenges. Smart grids, which use digital technology to manage and distribute electricity, rely on precise measurements of electricity flow, voltage, and frequency. These measurements are critical for balancing supply and demand, preventing outages, and maintaining grid stability. Metrology supports the development of advanced sensors and monitoring systems that provide real-time data on the performance of renewable energy sources and their interaction with the grid. Furthermore, accurate measurements are essential for assessing the efficiency and environmental impact of energy storage systems, such as batteries and supercapacitors, which are integral to the functioning of smart grids [32, 33].

3.5 Lifecycle Assessment

Metrology also plays a crucial role in the lifecycle assessment (LCA) of green technologies. LCA involves evaluating the environmental impact of a product or system from cradle to grave, including raw material extraction, manufacturing, usage, and disposal [34]. Accurate measurement of material usage, energy consumption, emissions, and waste generation is necessary to identify areas for improvement and to ensure that green technologies truly contribute to sustainability. This requires the development of standardized measurement protocols and robust data collection methods.

4 Sustainable Manufacturing and Circular Economy

Sustainable manufacturing and the circular economy are integral to reducing the environmental impact of production and consumption. Sustainable manufacturing aims to minimize waste and resource consumption while maximizing product life and recyclability. Metrology is fundamental in achieving these goals by providing precise measurements for quality control, process optimization, and material characterization. For example, in additive manufacturing (3D printing), accurate measurement of layer thickness, material composition, and structural integrity is crucial for producing high-quality and durable products. These approaches focus on minimizing waste, optimizing resource use, and extending the lifecycle of products, thereby contributing to a more sustainable future [5, 6, 35].

4.1 Sustainable Manufacturing

Sustainable manufacturing aims to reduce the environmental footprint of production processes while maintaining economic viability. This involves optimizing energy and material usage, minimizing waste, and ensuring the safety and health of workers. Key strategies include:

4.1.1 Resource Efficiency

Implementing techniques to use raw materials and energy more efficiently, thereby reducing costs and environmental impact. This includes the adoption of renewable energy sources and the implementation of energy-saving technologies.

4.1.2 Waste Minimization

Developing processes that generate less waste and promoting the reuse and recycling of materials within the production cycle. For example, lean manufacturing techniques aim to eliminate waste by streamlining production processes.

4.1.3 Green Design

Creating products with sustainability in mind, considering the entire lifecycle from raw material extraction to disposal. This involves using environmentally friendly materials, designing for disassembly, and ensuring products are durable and repairable.

4.1.4 Process Optimization

Utilizing advanced manufacturing technologies such as additive manufacturing (3D printing) and precision engineering to enhance efficiency and reduce material waste.

4.2 Circular Economy

The circular economy concept goes beyond sustainable manufacturing by emphasizing the continuous use of resources through a closed-loop system. In contrast to the traditional linear economy (take-make-dispose), the circular economy seeks to keep products, components, and materials in use for as long as possible. Key principles include the following.

4.2.1 Design for Longevity

Creating products that are durable, repairable, and upgradable, thus extending their useful life. This reduces the need for frequent replacements and conserves resources.

4.2.2 Reuse and Remanufacture

Encouraging the return of used products for refurbishment or remanufacturing, thereby restoring them to like-new condition. This not only saves materials but also reduces energy consumption compared to producing new items.

4.2.3 Recycling

Ensuring that materials from end-of-life products are recovered and reprocessed into new products. This requires developing efficient recycling technologies and systems for collecting and sorting waste.

4.2.4 Resource Recovery

Extracting valuable materials from waste streams to be reused in manufacturing processes. This includes the recovery of metals, plastics, and other materials from discarded products.

4.3 Metrological Challenges

Accurate and reliable measurements are crucial for both sustainable manufacturing and the circular economy. Metrology provides the tools necessary for quality control, process optimization, and material characterization. Challenges include the following.

(i) Ensuring the precise measurement of material properties and production parameters to maintain quality and efficiency.
(ii) Developing and harmonizing standards for measuring the quality and purity of recycled materials, ensuring they meet required specifications for reuse.
(iii) Advancing sensor technologies for real-time monitoring of manufacturing processes and environmental conditions, enabling immediate adjustments to reduce waste and improve efficiency.

5 Health and Safety in a Sustainable Future

As the world pursues sustainability, ensuring the health and safety of populations remains a priority. Metrology plays a critical role in monitoring environmental pollutants, food safety, and occupational hazards. For instance, accurate measurement of air and water quality is essential for assessing the impact of pollution on public health and implementing effective mitigation strategies. As the world intensifies efforts toward sustainability, ensuring the health and safety of populations becomes increasingly crucial. Metrology, the science of measurement, is integral to these efforts, providing the precision necessary to monitor environmental pollutants, ensure food safety, and manage occupational hazards effectively. Accurate and reliable measurements are foundational for assessing risks and implementing strategies to protect public health within the framework of sustainable development [36, 37] (Fig. 4).

5.1 *Environmental Pollutants and Public Health*

Environmental pollutants, such as air and water contaminants, pose significant threats to public health. Precise measurement of these pollutants is vital for understanding their sources, distribution, and impact. Advanced metrological techniques enable the detection of trace amounts of harmful substances, such as heavy metals, particulate matter, and volatile organic compounds. Continuous monitoring systems, equipped with high-precision sensors, provide real-time data on pollutant levels, helping to

Fig. 4 Metrological issues for health safety and sustainable future

identify pollution hotspots and trends over time. This data is essential for regulatory agencies to enforce environmental standards and for communities to take informed actions to reduce exposure and mitigate health risks [11, 12].

5.2 Food Safety and Sustainable Agriculture

Food safety is another critical aspect of health in a sustainable future. The increasing demand for sustainable agriculture practices necessitates accurate measurements of soil health, water usage, and crop quality. Metrology supports the monitoring of pesticide residues, heavy metals, and other contaminants in food products, ensuring they meet safety standards. Precision agriculture techniques, which rely on detailed measurements of environmental conditions, enable farmers to optimize resource use, reducing the need for chemical inputs and minimizing environmental impact. By improving the accuracy of these measurements, metrology helps enhance food security and safety while promoting sustainable farming practices [5].

5.3 Occupational Health and Safety

In the context of occupational health and safety, metrology is essential for monitoring workplace environments and ensuring compliance with safety regulations. Accurate measurement of factors such as air quality, noise levels, and exposure to hazardous substances is crucial for protecting workers' health. Sustainable industries must prioritize the health and safety of their workforce by implementing robust monitoring systems. For example, in manufacturing and construction, wearable sensors and smart devices can provide continuous data on workers' exposure to potential hazards, allowing for timely interventions and the prevention of occupational illnesses and injuries [10, 38–40].

5.4 Safety of New Materials and Technologies

The development and deployment of new materials and technologies for sustainability—such as bio-based plastics, advanced composites, and nanomaterials—require thorough safety assessments. Metrology provides the tools necessary for evaluating the toxicity, durability, and environmental impact of these materials [41, 42]. Rigorous testing protocols and precise measurements ensure that new products are safe for use and disposal, mitigating potential risks to human health and the environment. As innovation accelerates, maintaining high standards of safety through accurate measurement becomes ever more important.

6 International Collaboration and Standardization

Achieving a sustainable future requires global cooperation and the harmonization of measurement standards. The establishment of internationally accepted metrological standards is crucial for ensuring the comparability and reliability of data across borders. This facilitates the sharing of information and best practices, enabling countries to work together in addressing global challenges such as climate change and resource scarcity [9, 10, 42].

International collaboration and standardization in metrology are pivotal to addressing global challenges and ensuring a sustainable future. As the world faces interconnected issues such as climate change, resource scarcity, and environmental degradation, harmonized measurement standards and cooperative efforts are essential for effective action and decision-making.

6.1 The Need for International Standards

Global sustainability initiatives rely heavily on precise and consistent measurements. For instance, monitoring greenhouse gas emissions, assessing air and water quality, and tracking biodiversity loss requires data that are accurate and comparable across different regions and time periods. Without standardized measurement methods, data discrepancies can lead to misunderstandings, policy misalignments, and ineffective interventions [43].

International standards ensure that measurements taken in one country are compatible with those taken elsewhere, facilitating the comparison and aggregation of data. This harmonization is critical for global initiatives like the Paris Agreement, which depends on accurate emission inventories to track progress and set targets [44].

6.2 Key Organizations in Metrological Standardization

Several key organizations spearhead the development and promotion of international measurement standards:

International Bureau of Weights and Measures (BIPM): BIPM plays a central role in maintaining the International System of Units (SI) and coordinating global metrology efforts. It ensures that units of measurement are uniform and widely accepted, providing a foundation for international trade, scientific research, and regulatory compliance.

International Organization for Standardization (ISO): ISO develops voluntary, consensus-based international standards across various sectors. ISO standards cover environmental management, energy efficiency, and product quality, among others, contributing to sustainability by promoting best practices and reducing environmental impacts.

International Electrotechnical Commission (IEC): IEC focuses on standards for electrical, electronic, and related technologies. As renewable energy technologies like solar and wind power expand, IEC standards help ensure the safety, reliability, and efficiency of these systems.

6.3 Collaboration Mechanisms

Effective international collaboration in metrology involves several mechanisms:

Joint Research Projects: Countries and organizations collaborate on research projects to develop new measurement techniques and standards. These projects often address emerging challenges, such as the need for more precise climate data or the development of new materials for sustainable technologies.

Interlaboratory Comparisons: To ensure consistency and accuracy, laboratories around the world participate in interlaboratory comparisons, where they test their measurement capabilities against standardized references. This process helps identify discrepancies and improve measurement techniques.

Capacity Building and Knowledge Sharing: Developed countries and international organizations provide technical assistance, training, and resources to developing nations. This capacity building ensures that all countries have the necessary tools and expertise to participate in global metrology efforts.

6.4 Challenges and Future Directions

While significant progress has been made, challenges remain in achieving comprehensive international collaboration and standardization. Differences in technological capabilities, economic resources, and regulatory environments can hinder the implementation of uniform standards. Addressing these disparities requires sustained effort and investment in global metrological infrastructure. Looking ahead, the focus will likely shift toward integrating new technologies such as artificial intelligence and the Internet of Things (IoT) into metrological practices. These advancements can enhance data collection, analysis, and dissemination, further supporting global sustainability efforts.

7 Conclusion

Metrology is at the forefront of addressing the challenges posed by the transition to a sustainable future. From environmental monitoring and renewable energy to sustainable manufacturing and public health, accurate and reliable measurements are essential for achieving sustainability goals. The development of advanced metrological techniques, the establishment of robust calibration standards, and international collaboration are key to overcoming these challenges. As we move toward a more sustainable world, the role of metrology will continue to expand, providing the foundation for informed decision-making and innovative solutions to global challenges.

International collaboration and standardization in metrology are essential for addressing global sustainability challenges. Precise and consistent measurements are crucial for monitoring environmental changes, tracking progress in climate agreements, and ensuring the reliability of green technologies. Organizations like the International Bureau of Weights and Measures (BIPM) and the International Organization for Standardization (ISO) play a vital role in developing and maintaining universally accepted measurement standards. These standards enable accurate data comparison and aggregation across different regions, which is critical for effective global initiatives.

Key mechanisms for international collaboration include joint research projects, interlaboratory comparisons, and capacity building. Joint research projects help develop new measurement techniques to address emerging challenges, while interlaboratory comparisons ensure measurement consistency and accuracy worldwide. Capacity-building efforts, particularly in developing nations, are crucial for ensuring all countries have the tools and expertise needed to participate in global metrology efforts. These collaborative activities foster a shared understanding and implementation of measurement standards, promoting global sustainability.

Despite significant progress, challenges remain, such as disparities in technological capabilities and resources among countries. Addressing these challenges requires sustained investment in global metrological infrastructure and continued efforts to harmonize measurement standards. The integration of advanced technologies like artificial intelligence and the Internet of Things (IoT) into metrological practices promises to enhance data collection and analysis, further supporting global sustainability goals. Overall, international collaboration and standardization in metrology are indispensable for informed decision-making and coordinated action toward a sustainable future.

References

1. Climate change 'biggest threat modern humans have ever faced', world-renowned naturalist tells security council, calls for greater global cooperation, a UN press note during (2021)

2. Tamador Salih (2024), Metrology in Climate Change: Ensuring Accurate Measurements for a Sustainable Future, https://www.linkedin.com/pulse/metrology-climate-change-ensuring-accurate-future-tamador-salih-1yphf/
3. Shivanna KR (2022) Climate change and its impact on biodiversity and human welfare. Proc Indian Natl Sci Acad. 88(2):160–71. https://www.ncbi.nlm.nih.gov/pmc/articles/PMC9058818/
4. Adeniyi Kehinde Adeleke, Emmanuel Chigozie Ani, Kehinde Andrew Olu-lawal, Oladiran Kayode Olajiga, Danny Jose Portillo Montero (2024) Future of precision manufacturing: Integrating advanced metrology and intelligent monitoring for process optimization. Int J Sci Res Arch, 11(01), 2346–2355. https://doi.org/10.30574/ijsra.2024.11.1.0335, https://ijsra.net/content/future-precision-manufacturing-integrating-advanced-metrology-and-intelligent-monitoring
5. Yadav S, Rab S, Wan M, Bhatnagar A, Jain A, Achanta VG (2024) Role of conformity assessment and metrology in the circular economy for sustainable growth. In: Bhatnagar A, Yadav S, Achanta V, Harmes-Liedtke U, Rab S (eds) Handb Qual Syst, Accredit Conform Assess. Springer, Singapore. https://doi.org/10.1007/978-981-99-4637-2_19-1
6. Diaz LF (2017) Waste management in developing countries and the circular economy. Waste Manag Res 35(1):1–2
7. Yadav S, Rab S, Wan M (2024) Metrology and sustainability in industry 6.0: navigating a new paradigm. In: Bhatnagar A, Yadav S, Achanta V, Harmes-Liedtke U, Rab S (eds) Handb Qual Syst, Accredit Conform Assess. Springer, Singapore. https://doi.org/10.1007/978-981-99-4637-2_64-1
8. Chourasia S, Tyagi A, Pandey SM et al (2022) Sustainability of Industry 6.0 in global perspective: benefits and challenges. Mapan 37:443–452
9. Aswal DK (2020) Introduction: metrology for all people for all time. In: Aswal DK (eds) Metrol Incl Growth India. Springer, Singapore. https://doi.org/10.1007/978-981-15-8872-3_1
10. Yadav S, Rab S, Jaiswal SK, Kumar A, Aswal DK (2023) Industrial metrology. In: Aswal DK, Yadav S, Takatsuji T, Rachakonda P, Kumar H (eds) Handb Metrol Applications Springer, Singapore. https://doi.org/10.1007/978-981-99-2074-7_34
11. Sharma C, Mandal TK, Singh S, Gupta G, Kulshrestha MJ, Johri P, Ranjan A, Upadhayaya AK, Das RM, Soni D, Mishra SK, Muthusamy SK, Sharma SK, Singh P, Aggarwal SG, Radhakrishnan SR, Kumar M (2020) Metrology for atmospheric environment—part I: atmospheric constituents. In: Aswal DK (ed) Metrology for inclusive growth of India. Springer, Singapore, pp 639–689. https://doi.org/10.1007/978-981-15-8872-3_13
12. Sharma C et al (2020). Metrology for atmospheric environment. In: Aswal DK (eds) Metrol Incl Growth India. Springer, Singapore. https://doi.org/10.1007/978-981-15-8872-3_14
13. Rajesh Gajjala S, Agrawal VV, Aswal DK (2020) Biomedical metrology: role in nation's healthcare sector. In: Aswal DK (eds) Metrol Incl Growth India. Springer, Singapore. https://doi.org/10.1007/978-981-15-8872-3_15
14. Rab S, Wan M, Yadav S (2022) Let's get digital. Nat Phys 18:960. https://doi.org/10.1038/s41567-022-01707-8
15. Brown RJC (2021) Measuring measurement—What is metrology and why does it matter?, Measurement. 168:108408, https://doi.org/10.1016/j.measurement.2020.108408
16. Garg N, Rab S, Varshney A, Jaiswal SK, Yadav S (2021) Significance and implications of digital transformation in metrology in India. Meas Sens 18:100248. https://doi.org/10.1016/j.measen.2021.100248
17. Rab S, Wan M, Sharma RK, Kumar L, Zafer A, Saeed K, Yadav S (2023) Digital avatar of metrology. MAPAN J Metrol Soc India. https://doi.org/10.1007/s12647-023-00641-1
18. Varshney A, Garg N, Nagla KS, Nair TS, Jaiswal SK, Yadav S, Aswal DK (2021) Challenges in sensors technology for industry 4.0 for futuristic metrological applications. Mapan 36(2):215–226
19. Overview of Greenhouse Gases, A blog post accessed during July, 2024, https://www.epa.gov/ghgemissions/overview-greenhouse-gases.

20. Andrew F (2023) Feldman, Zhen Zhang, Yasuko Yoshida, Abhishek Chatterjee, and Benjamin Poulter, Satellite instruments like NASA's Orbiting Carbon Observatory-2 (OCO-2) offer global coverage and enable the detection of even minor changes in atmospheric CO2 levels. Atmos Chem Phys 23:1545–1563. https://doi.org/10.5194/acp-23-1545-2023
21. Sandven S, Spreen G, Heygster G et al (2023) Sea Ice Remote sensing—recent developments in methods and climate data sets. Surv Geophys 44:1653–1689. https://doi.org/10.1007/s10712-023-09781-0
22. https://www.bipm.org/en/.
23. https://wmo.int/
24. Krishna Kumar Jaiswal (2022) Chandrama Roy Chowdhury. Deepti Yadav, Ravikant Verma, Swapnamoy Dutta, Km Smriti Jaiswal, SangmeshB, Karthik Selva Kumar Karuppasamy, Renewable and sustainable clean energy development and impact on social, economic, and environmental health, Energy Nexus 7:100118. https://doi.org/10.1016/j.nexus.2022.100118
25. https://www.nplindia.org/index.php/science-technology/advanced-materials-and-device-metrology-division/photovoltaic-metrology-section/
26. Emilio Fernandez Lisbona, Calibration, Testing and Monitoring of Space Solar Cells, Editor(s): Tom Markvart, Luis Castañer, in Practical Handbook of Photovoltaics, Elsevier Science, 2003, Pages 825–853, https://doi.org/10.1016/B978-185617390-2/50036-2.
27. Porté-Agel F, Bastankhah M, Shamsoddin S (2020) Wind-Turbine and wind-farm flows: a review. Bound-Layer Meteorol 174:1–59. https://doi.org/10.1007/s10546-019-00473-0
28. Weidinger P, Foyer G, Kock S, Gnauert J, Kumme R (2019) Calibration of torque measurement under constant rotation in a wind turbine test bench. J Sens Sens Syst 8:149–159. https://doi.org/10.5194/jsss-8-149-2019
29. Bo Jing, Zheng Qian, Yan Pei, Lizhong Zhang, Tingyi Yang (2020) Improving wind turbine efficiency through detection and calibration of yaw misalignment, Renewable Energy, 160, 1217–1227, https://doi.org/10.1016/j.renene.2020.07.063
30. Sivasubramanian Manikandan, Sundaram Vickram, Ranjna Sirohi, Ramasamy Subbaiya, Radhakrishnan Yedhu Krishnan, Natchimuthu Karmegam, C. Sumathijones, Rajinikanth Rajagopal, Soon Woong Chang, Balasubramani Ravindran, Mukesh Kumar Awasthi (2023) Critical . review of biochemical pathways to transformation of waste and biomass into bioenergy. Bioresour Technol, 372, 128679
31. Moaaz Shehaba, Camelia Stratulatc, Kemal Ozcand, Aylin Bozteped, Fatma Coskund, Feyzanur Senturkd Alper Isleyend, Edwin Zondervanb, Kai Moshammera (2022) Improved metrological methodology to address the challenges associated with the determination of biofuels calorific value by bomb calorimeter, Chem Eng Trans, 92, https://www.cetjournal.it/cet/22/92/073.pdf
32. Chen Z (2023) Ali Moradi Amani, Xinghuo Yu and Mahdi Jalili, Control and optimisation of power grids using smart meter data: a review. Sens 23(4):2118. https://doi.org/10.3390/s23042118
33. Uddin M, Mo H, Dong D, Elsawah S, Zhu J, Guerrero JM (2023) Microgrids: A review, outstanding issues and future trends. Energ Strat Rev 49:101127. https://doi.org/10.1016/j.esr.2023.101127
34. Tamoor M, Samak NA, Yang M, Xing J (2022) The Cradle-to-Cradle life cycle assessment of polyethylene terephthalate: environmental perspective. Mol 27:1599. https://doi.org/10.3390/molecules27051599
35. Kumar H, Yadav S (2024) Additive and good manufacturing practices in conformity assessment. In: Bhatnagar A, Yadav S, Achanta V, Harmes-Liedtke U, Rab S (eds) Handb Qual Syst, Accredit Conform Assess. Springer, Singapore. https://doi.org/10.1007/978-981-99-4637-2_53-1
36. Durazzo A, Souto EB, Lombardi-Boccia G, Santini A, Lucarini M (2021) Metrology, agriculture, and food: literature quantitative analysis. Agric 11(9):889. https://doi.org/10.3390/agriculture11090889
37. Sorbo A, Pucci E, Nobili C, Taglieri I, Passeri D, Zoan C (2022) Food safety assessment: overview of metrological issues and regulatory aspects in the European Union. Sep 9(2):53. https://doi.org/10.3390/separations9020053

38. Brocal F, Sebastian MA, Gonzalez C (2016) Classification proposal of metrological techniques in occupational safety and health, Proceedings of the 26th DAAAM International Symposium, pp 0648–0655, Katalinic B (Ed), Published by DAAAM International, ISBN 978–3–902734–07–5, ISSN 1726–9679, Vienna, Austria, https://doi.org/10.2507/26th.daaam.proceedings.088
39. Rana BK, Yadav D, Yadav S (2024) Strategic association for quality excellence: coordination between GMP, GLP, GCP, Metrology, QC, and QA. In: Bhatnagar A, Yadav S, Achanta V, Harmes-Liedtke U, Rab S (eds) Handb Qual Syst, Accredit Conform Assess. Springer, Singapore. https://doi.org/10.1007/978-981-99-4637-2_29-1
40. Feller E, Jaoui Y (2024) Intersection between requirements of OECD-GLP, GCLP, and ISO/IEC 17025, ISO 15189 Standards. In: Bhatnagar A, Yadav S, Achanta V, Harmes-Liedtke U, Rab S (eds) Handb Qual Syst, Accredit Conform Assess. Springer, Singapore. https://doi.org/10.1007/978-981-99-4637-2_63-2
41. Pollard A, Castro, F, Edwards G, Gee M, Giannis S, Kitchen S, Williams J (2024) Adv Mater Metrol Strat, https://doi.org/10.47120/npl.9944
42. Rab S, Wan M, Yadav S (2023) International and national metrology. In: Aswal DK, Yadav S, Takatsuji T, Rachakonda P, Kumar H (eds) Handb Metrol Appl. Springer, Singapore. https://doi.org/10.1007/978-981-99-2074-7_2
43. Yadav BK (2024) Impact of regulation and conformity assessment procedures on global trade. In: Bhatnagar A, Yadav S, Achanta V, Harmes-Liedtke U, Rab S (eds) Handb Qual Syst, Accredit Conform Assess. Springer, Singapore. https://doi.org/10.1007/978-981-99-4637-2_14-1
44. Singh T (2024) Good regulatory practices and its evolving relevance in conformity assessment. In: Bhatnagar A, Yadav S, Achanta V, Harmes-Liedtke U, Rab S (eds) Handb Qual Syst, Accredit Conform Assess. Springer, Singapore. https://doi.org/10.1007/978-981-99-4637-2_16-1

A Study on the Effect of Crop-Residue Burning (CRB) on Air Quality Over the Megacity Delhi During Winter Season

Priya Dwivedi, S. R. Radhakrishnan, Deepak Kumar, and S. K. Sharma

Abstract Crop-Residue Burning (CRB) over northern India is a serious problem for both human health and for the air quality during the rabi and kharif seasons. The data obtained from two Continuous Ambient Air Quality Monitoring Stations (CAAQMS) of Central Pollution Control Board (CPCB) installed at DMS (Delhi Milk Scheme)-Shadipur, and NSUT (Netaji Subhas University of Technology)—Dwarka is used for the study of variation of PM_{10}, $PM_{2.5}$, NO_2, SO_2 & CO over Delhi during the crop residue burning period in 2023. A detailed analysis of the above parameters 'Pre—Burning' (October 1 to October 30, 2023) and 'Burning' (November 1 to December 5, 2023) has been carried out. The results showed that PM_{10}, $PM_{2.5,}$ NO_2, SO_2 and CO from October to November indicated a Significant percentage changes in air pollution levels over Delhi, value of PM_{10}, $PM_{2.5,}$ NO_2, SO_2 and CO were found to be increased by nearly, 170%, 130%, 43%, 348%, and 76% respectively over DMS, and 180%, 230%, 237%, 111.45%, and 335% respectively (all values are in ug/ m^3) over NSIT. Pollutant concentrations are influenced by weather (meteorological parameters) patterns in both seasons (Rabi & Kharif). Air pollution is made worse by regular dust storms during October–November and also due to fireworks celebrations of Diwali and Dusshera. There is an inverse relationship between (relative humidity) RH and (air temperature) AT for PM_{10} & $PM_{2.5}$. High wind speed and dominant wind direction in this season promote PM_{10}, $PM_{2.5}$, NO_2, CO and SO_2 dispersion on a large scale which leads transboundary movement of gaseous and dust pollutant. MODIS Fire count data's backward trajectories depict air mass movement from Haryana and Punjab to Delhi via wind movement which deteriorated air quality during the study period.

Keywords Delhi Air quality · CRB · National Air quality index (NAQI)

P. Dwivedi · S. R. Radhakrishnan (✉)
Environmental Sciences and Biomedical Metrology Division, CSIR-National Physical Laboratory, New Delhi, India
e-mail: rkrishnan.nplindia@csir.res.in

Academy of Scientific and Innovative Research (AcSIR), Ghaziabad, India

D. Kumar · S. K. Sharma
Air Lab, Central Pollution Control Board (CPCB), New Delhi, India

S. Yadav et al. (eds.), *Advances in Metrology*, Lecture Notes in Mechanical Engineering,
https://doi.org/10.1007/978-981-96-6418-4_14

1 Introduction

Crop burning in Haryana and Punjab emerged as a significant issue in the late 1980s and early 1990s, coinciding with the widespread use of mechanized combine harvesters. These machines left behind substantial crop residues after harvesting rice and wheat. To clear the fields quickly and cost-effectively for the next planting season, farmers traditionally burned these residues. Although initially a local problem, the practice expanded considerably by the early 2000s due to increased farm mechanization and the need to manage larger harvests efficiently [1–3]. This has since become a major contributor to the air pollution and smog crisis in northern India, particularly during winter months. Crop residue burning in Punjab and Haryana significantly affects air quality in Delhi by releasing large amounts of smoke, particulate matter, and harmful gases like carbon monoxide and sulfur dioxide. These pollutants can travel long distances, degrading air quality in Delhi, causing health issues such as respiratory problems, asthma, and bronchitis. Additionally, the smoke can reduce visibility, disrupt transportation, and impact agricultural productivity (Fig. 1) [2, 3]. The impact of crop burning in Punjab and Haryana on Delhi's air quality is mainly influenced by geographical and meteorological factors. Delhi is located southeast and downwind of these states, where most of the crop burning occurs after the harvest season. As a result, the pollutants from stubble burning in Punjab and Haryana are transported by prevailing wind patterns towards Delhi [3, 4]. During the post-monsoon months of October and November, when crop burning is at its peak, weather conditions typically include stable atmospheric layers and lower wind speeds. These factors can cause pollutants to become trapped and accumulate in Delhi's air basin. Furthermore, temperature inversions, where a layer of warm air traps cooler air near the ground, can intensify pollution levels by hindering the vertical mixing and dispersion of pollutants [4, 5]. Delhi's unique topography and urban heat island effect also contribute to the concentration of pollutants. The city's built environment and dense population create localized weather patterns that can trap pollutants within the urban area, leading to prolonged exposure and higher concentrations of harmful substances in the air [5, 6]. Overall, the combination of geographic alignment, prevailing wind directions, meteorological conditions, and urban characteristics make Delhi particularly susceptible to the impacts of crop burning from Punjab and Haryana (Fig. 2). Addressing this issue requires comprehensive regional and transboundary strategies to mitigate emissions at the source and improve air quality management across affected areas [6, 7].

Delhi is an ideal location for studying the impact of crop burning residue due to its geographical proximity to major agricultural states like Punjab, Haryana, and Uttar Pradesh, where extensive crop residue burning occurs after harvest seasons (Fig. 1). The geographical location of Punjab, Haryana, and Delhi shown in Fig. 2 and location map of selected stations in Delhi [Fig. 2] are drawn using the QGIS website, which is a publicly accessible platform for drawing the geographical map of earth (https://www.qgis.org/). This practice significantly contributes to Delhi's persistent air quality issues, particularly during the winter months when smoke and

Fig. 1 Burning of rice residues, a prevalent practice in northwest India

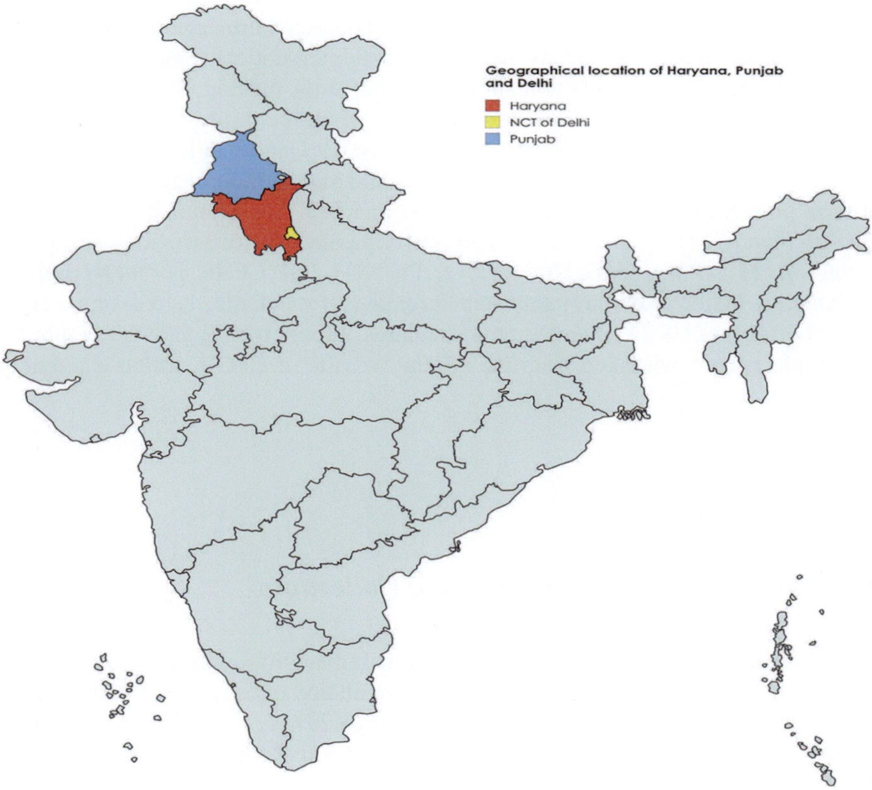

Fig. 2 Geographical location of Haryana, Punjab and Delhi

Table 1 National ambient air quality standards

Pollutant	Unit	NO_2	SO_2	PM_{10}	$PM_{2.5}$	Benzene	NH_3
Indian Std.(24hrly)	$\mu g/m^3$	80	80	100	60		400
Indian Std (Yearly)	$\mu g/m^3$	40	50	60	40	5	100
Pollutant	Unit	CO	O_3				
Indian Std (1hrly)	$\mu g/m^3$	4000	180				
Indian Std (8hrly)	$\mu g/m^3$	2000	100				

particulate matter from burning fields engulfs the city. The resulting air pollution poses serious health risks to Delhi's population, exacerbating respiratory illnesses and increasing the incidence of cardiovascular diseases. Furthermore, Delhi serves as a critical focal point for policy discussions and interventions aimed at mitigating crop burning and improving urban air quality standards, making it a crucial area for research to understand the direct consequences of agricultural practices on public health and environmental sustainability [7, 8]. The Air Quality Index (AQI) typically consolidates individual pollutant data into a single index, traditionally based on five criteria pollutants: PM_{10}, $PM_{2.5}$, SO_2, NO_2, and CO. The National Ambient Air Quality Standards (NAAQS) introduced twelve parameters for assessing air quality, including PM_{10}, $PM_{2.5}$, SO_2, NO_2, CO, O_3, Pb, NH3, BaP, C6H6, as, and Ni (https://cpcb.nic.in). While only four parameters have annual standards, the remaining eight have short-term (1/8/24 h) and annual standards, except for O_3 and CO (Table 1). The Table 1 has been taken from the official website of CPCB, which is publicly accessible.

2 Methodology

2.1 Monitoring Location and Data Collection

Delhi, the National Capital Territory, like most Indian cities, has a mixed land-use area. Delhi has an area of 1,483 sq. km and a population of nearly 16.7 million, with one of the highest population densities in India at 11,297 people per square kilometer [8–10]. The increase in population and economic activity has led to increased air pollution in Delhi. The main sources of air pollution in Delhi are Vehicular source, Industrial sources, Domestic sources etc.

It is estimated that road traffic in Delhi contributes approximately 70% of air pollution in urban areas. The vehicular population in Delhi is nearly 70 Lakhs (67% Two Wheelers, 24% Four Wheelers, 3% Three Wheelers, 5% Commercial Vehicles, and 1% Buses) [11–13].

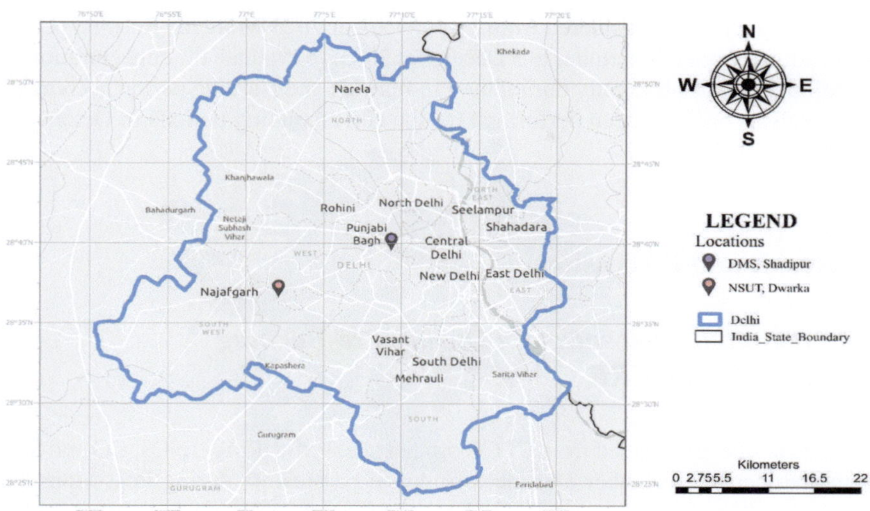

Fig. 3 Location map of selected CAAQMS stations in Delhi

Central Pollution Control Board (CPCB) has established Continuous Ambient Air Monitoring Systems in various cities for Ambient Atmospheric Air Data analysis, including in Delhi. We selected two different stations in Delhi for the current studies, shown in (Fig. 3) are as follows:

(a) DMS (Delhi Milk Scheme) at Shadipur ($28^0 39'30.91''$N, $77^0 09'21.93''$E) is located in the vicinity of DMS. DMS Station is surrounded by Kirti Nagar and Najafgarh Road Industrial Area. Parameters which are measured at DMS are: Carbon Monoxide (CO), Respirable Suspended Particulate Matter ($PM_{2.5}$ and PM_{10}), Oxides of Nitrogen (NO_X), Ozone (O_3), Sulphur Dioxide (SO_2), Benzene-Toluene-Xylene (BTX), Ambient Temperature, Relative Humidity, Wind Speed, Wind Direction, Vertical Wind Speed, Ambient Pressure and Solar Radiation.

(b) NSUT (Netaji Subhas University of Technology) at Dwarka ($28^0 36'37.14''$N, $77^0 02'6.66''$E). CAAQMS at NSIT, Dwarka is located at $28^0 36'37.14''$N, $77^0 02'6.66''$E. It has a mixed land use area. Parameters which are measured at NSIT are: Carbon Monoxide (CO), Respirable Suspended Particulate Matter ($PM_{2.5}$), Oxides of Nitrogen (NO_X), Ozone (O_3), Sulphur Dioxide (SO_2), Benzene-Toluene-Xylene (BTX), Ambient Temperature, Relative Humidity, Wind Speed, Wind Direction, Vertical Wind Speed, Ambient Pressure and Solar Radiation.

Above all two stations are mixed land-use areas with residential and industrial sites in their vicinity. Data from the Continuous Ambient Air Quality Monitoring (CAAQM) network has been collected and analyzed for the current study obtained from the CPCB portal. The study period is divided into Pre-burning and Burning.

Pre-burning phase is considered from 1−31 October 2023 while the Burning period is considered from 1 November to 5 December 2023.To evaluate the concentration of air pollutant in the air quality during the Pre-burning and Burning period, a comparative analysis has also been performed for the corresponding periods in October to December 2023.

3 Result and Discussion

3.1 Pollutant Concentration During Pre-Burning and During CRB

The data has been collected from CPCB official site, PM_{10}, $PM_{2.5}$, NO_2, CO and SO_2 from October to December and meteorological parameters during Post-Monsoon and winter, (October-December2023).The monthly data obtained from two selected CAAQMS sites at Delhi during October 2023 to December 2023 is studied to understand the effect of CRB over Delhi air quality.The data from DMS and NSIT monitoring stations illustrate significant variations in air pollutant levels (Tables 2, 3, 4, 5, 6) (Figs. 4, 5, 6, 7, 8), specifically PM_{10}, $PM_{2.5}$, NO_2, SO_2, and CO, across three months: October, November, and December. In October, DMS recorded PM_{10}levels at 250.67 ± 89.70 μg/m^3, $PM_{2.5}$at 135.91 ± 57.98 μg/m^3, NO_2 at 63.77 ± 57.98 μg/m^3, SO_2 at 9.09 ± 1.68 μg/m^3, and COat 1672.55 ± 563.98 μg/m^3. NSIT, during the same period, observed slightly lower levels of PM_{10}243.57 ± 55.13 μg/m^3, $PM_{2.5}$114.61 ± 26.08 μg/m^3, NO2 28.14 ± 7.63 μg/m^3, SO2 11.08 ± 1.30 μg/m^3, and CO 758.70 ± 120.48 μg/m^3. From 1st November to 5th December 2023, the pollution levels increased notably at both stations. DMS reported PM_{10} rising to 678 ± 136.32 μg/m^3, $PM_{2.5}$to 313.57 ± 81.44 μg/m^3, NO_2 to 91.53 ± 81.44 μg/m^3, SO2 to 40.73 ± 3.43 μg/m^3, and CO to 2958.09 ± 290.37 μg/m^3. Similarly, NSIT recorded PM_{10} at 683.81 ± 114.62 μg/m^3, $PM_{2.5}$at 390.28 ± 64.04 μg/m^3, NO_2 at 95.11 ± 11.51 μg/m^3, SO_2 at 23.44 ± 4.82 μg/m^3, and CO at 3303.84 ± 871.45 μg/m^3.The number of fires recorded during the burning season in Punjab and Haryana provides compelling evidence supporting the deterioration of air quality in Delhi these fires significantly contribute to the increased levels of PM_{10}, $PM_{2.5}$, NO_2, SO_2, and CO observed during November and December [13, 14]. The heightened concentrations of these pollutants during this period exacerbate air pollution in Delhi, posing serious health risks to its residents [15] (Fig. 9).

The practice of crop burning in neighbouring states like Haryana and Punjab significantly deteriorates air quality in Delhi during the winter months. This conclusion is supported by the observed increases in pollutants such as PM_{10}, $PM_{2.5}$, NO_2, SO_2, and CO during the burning period compared to pre-burning months. These pollutants pose serious health risks to the population of Delhi, including respiratory and cardiovascular issues. Effective measures to curb agricultural burning, coupled

Table 2 PM_{10} (μg/m^3) over DMS, and NSUT Delhi CAAQMS Stations during the study period

Station	Period	Average PM_{10} concentration (μg/m^3)	Percentage change
DMS	Pre-CRB (1st -31st October)	250.67 ± 89.70	170.46
	During CRB (1stNovember–5th December)	678.0 ± 136.32	
NSIT	Pre-CRB (1st–31st October)	243.57 ± 55.13	180.75
	During CRB (1stNovember–5th December)	683.81 ± 114.62	

* CRB-Crop Residue Burning

Table 3 $PM_{2.5}$ (μg/m^3) over DMS, and NSUT Delhi CAAQMS Stations during the study period

Station	Period	Average $PM_{2.5}$ concentration (μg/m^3)	Percentage change
DMS	Pre-CRB (1st -31st October)	135.91 ± 57.98	130.75
	During CRB (1stNovember–5th December)	313.57 ± 81.44	
NSIT	Pre-CRB (1st–31st October)	114.61 ± 26.08	240.53
	During CRB (1stNovember–5th December)	390.28 ± 64.04	

* CRB-Crop Residue Burning

Table 4 NO_2 (μg/m^3) over DMS, and NSUT Delhi CAAQMS Stations during the study period

Station	Period	Average NO_2 concentration (μg/m^3)	Percentage change
DMS	Pre-CRB (1st–31st October)	63.77 ± 9.03	43.52
	During CRB (1stNovember–5th December)	91.53 ± 8.49	
NSIT	Pre-CRB (1st–31st October)	28 .14 ± 7.63	237.95
	During CRB (1stNovember–5th December)	95.11 ± 11.51	

* CRB-Crop Residue Burning

Table 5 SO_2 (μg/m^3) over DMS, and NSUT Delhi CAAQMS Stations during the study period

Station	Period	Average NO_2 concentration (μg/m^3)	Percentage change
DMS	Pre-CRB (1st–31st October)	9.09 ± 1.68	348.13
	During CRB (1stNovember–5th December)	40.73 ± 3.43	
NSIT	Pre-CRB (1st–31st October)	11.08 ± 1.30	111.45
	During CRB (1stNovember–5th December)	23.44 ± 4.82	

* CRB-Crop Residue Burning

Table 6 CO (μg/m³) over DMS, and NSUT Delhi CAAQMS Stations during the study period

Station	Period	Average NO_2 concentration (μg/m³)	Percentage change
DMS	Pre-CRB (1st–31st October)	1672.55 ± 563.98	76.86
	During CRB (1stNovember–5th December)	2958.09 ± 290.37	
NSIT	Pre-CRB (1st–31st October)	758.70 ± 120.48	335.45
	During CRB (1stNovember–5th December)	3303.84 ± 871.45	

* CRB-Crop Residue Burning

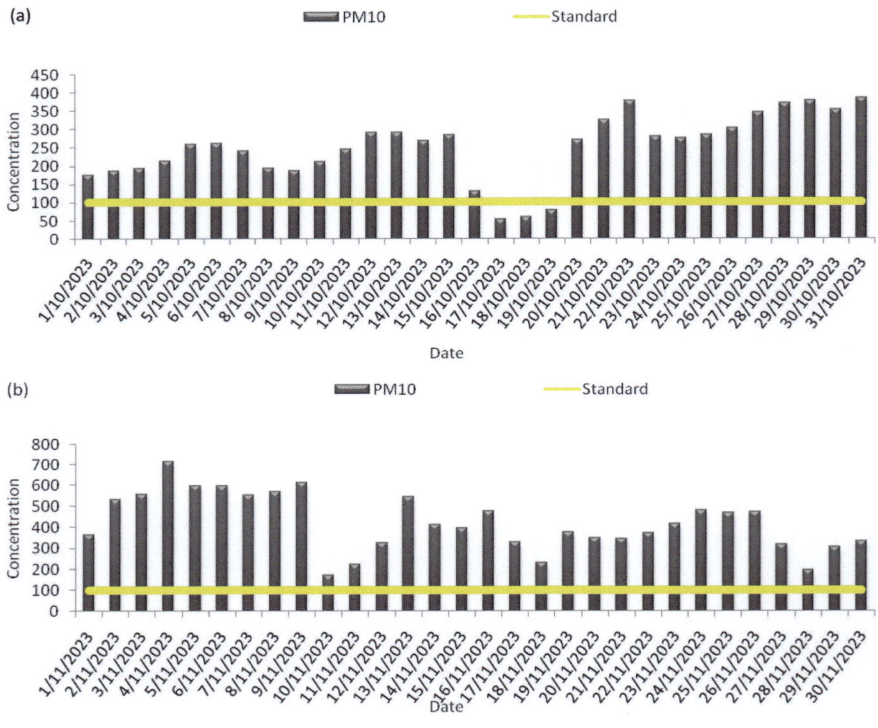

Fig. 4 PM_{10} concentration during **a** October to **b** November over Delhi during Pre- burning and Burning period

with enhanced regional air quality management strategies, are urgently needed to mitigate these impacts and safeguard public health in the national capital region.

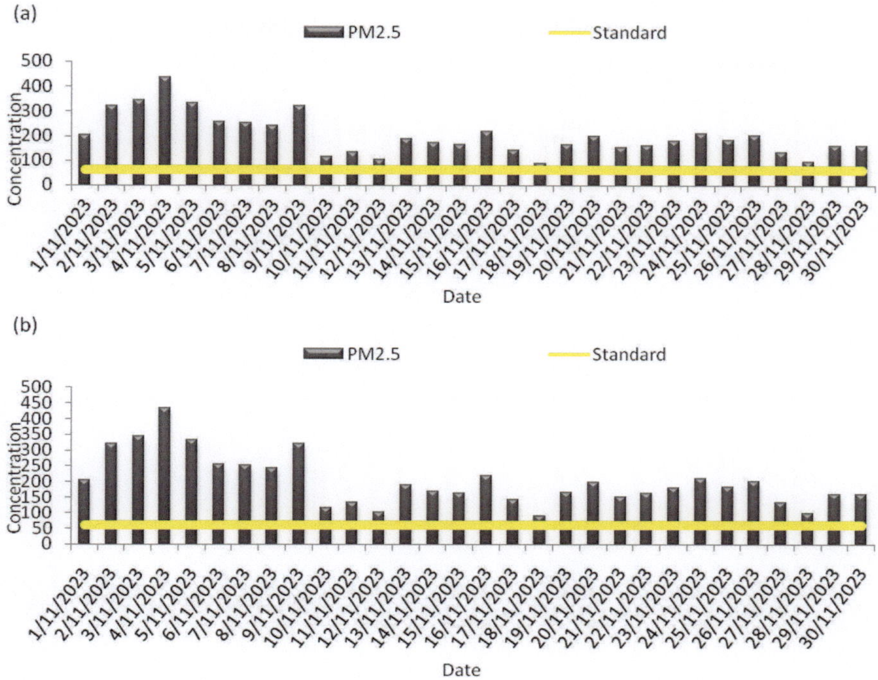

Fig. 5 $PM_{2.5}$ concentrations during **a** October to **b** November over Delhi during Pre- burning and Burning period

3.2 *Wind Movement During CRB*

During November and December in Delhi, the prevailing wind direction typically comes from the northwest or north-northwest. This wind pattern is influenced by several meteorological factors, including the movement of weather systems such as western disturbances (extra-tropical storms originating in the Mediterranean region and traveling eastwards towards South Asia) and the overall winter weather patterns over the Indian subcontinent [14, 16]. These winds bring cold air from the Himalayan region towards the plains of North India, including Delhi, and can also carry pollutants from agricultural crop burning in neighbouring states like Haryana and Punjab, contributing to the winter smog and air pollution in Delhi (Fig. 10).

The wind rose graph (Fig. 10) illustrates the prevailing wind direction patterns observed in Delhi during the months of October to December month of selected stations DMS and NSIT.

1. Wind Directions: The wind rose graph shows the distribution of wind directions in degrees, with 0° representing north and increasing clockwise around the compass rose.

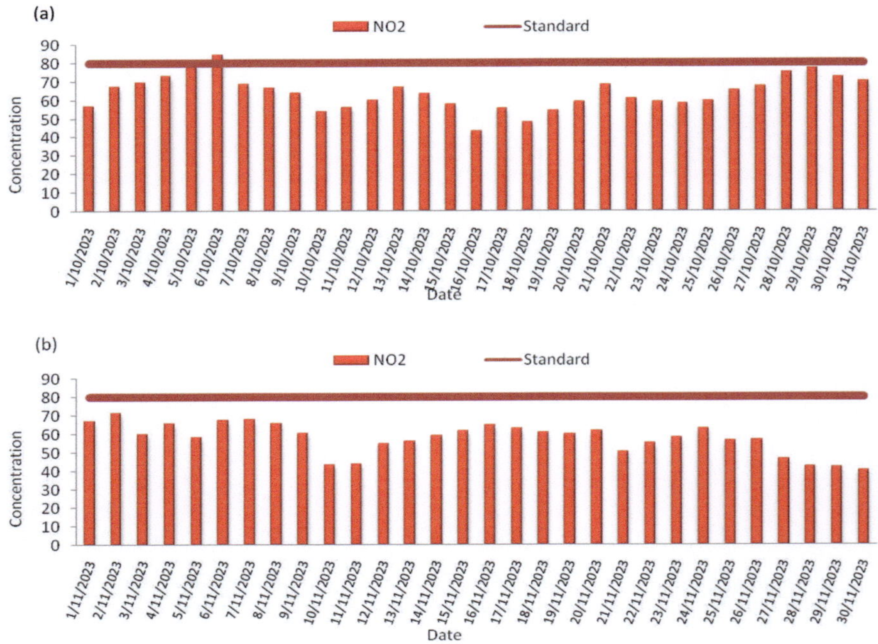

Fig. 6 NO_2 concentrations during **a** October to **b** November over Delhi during Pre- burning and Burning

2. Prevailing Wind Direction: During November and December, the prevailing wind in Delhi predominantly originates from the northwest (NW) and north-northwest (NNW) sectors of the compass (approximately 315° to 45°).
3. Meteorological Influences: These wind patterns are influenced by various meteorological factors, including the movement of weather systems such as western disturbances originating in the Mediterranean region and traveling eastwards towards South Asia. These systems bring cold air from the Himalayan region towards the plains of North India, affecting Delhi's weather patterns [16].

Pollutant Transport: Northwest and north-northwest winds can carry pollutants from agricultural crop burning in neighbouring states such as Haryana and Punjab. This transport contributes significantly to the winter smog and air pollution observed in Delhi during this period [14, 16].

3.3 Air Quality During CRB

Meteorological Factors: During the post-monsoon period, prevailing wind patterns often carry these emissions from Punjab and Haryana towards Delhi and the National

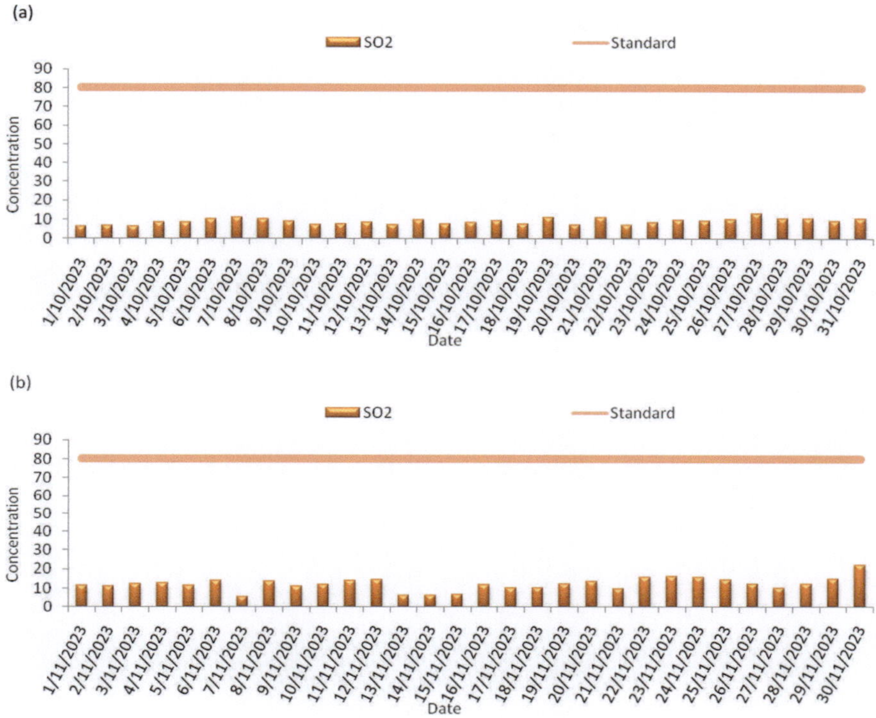

Fig. 7 SO_2 concentrations during **a** October to **b** November over Delhi during Pre- burning and Burning

Capital Region (NCR). This transport mechanism is crucial in determining the extent to which Delhi experiences pollution episodes originating from crop residue burning [17, 18]. The burning of crop residues releases carbonaceous aerosols (black carbon and brown carbon) which can affect regional climate dynamics [19–21]. These aerosols absorb sunlight, leading to localized warming effects and altering regional weather patterns. Impact on Delhi's Air Quality: studies have shown that Delhi's air quality deteriorates significantly during and after the crop residue burning season. The influx of $PM_{2.5}$ and PM_{10} particles increases pollution levels in Delhi, contributing to haze, reduced visibility, and adverse health effects for residents[22] (Fig. 11).

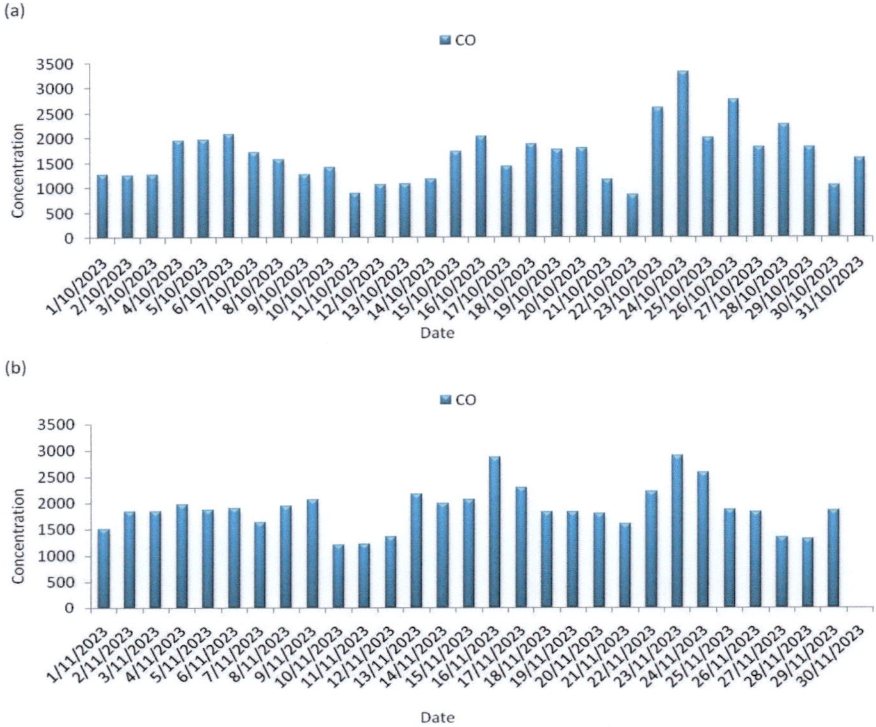

Fig. 8 CO concentrations during **a** October to **b** November over Delhi during Pre- burning and Burning

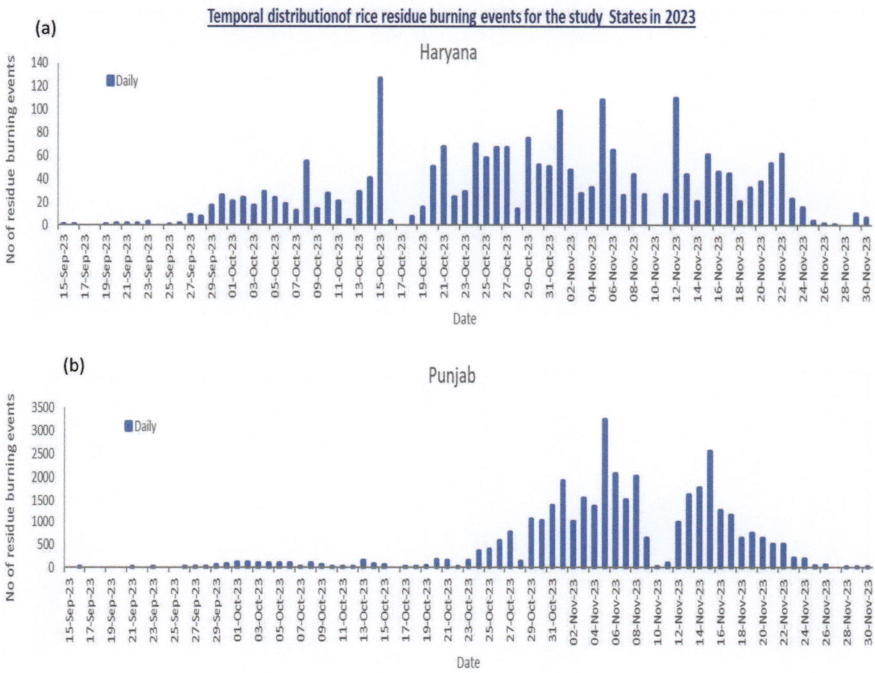

Fig. 9 Represent the burning events at **a** Haryana and **b** Punjab

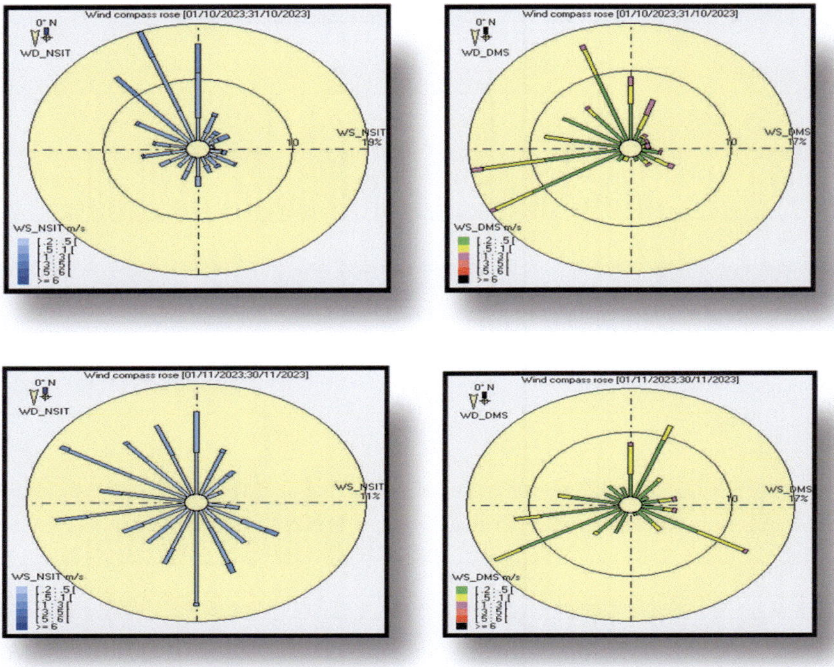

Fig. 10 Represents the wind direction during the crop residue burning period in Delhi

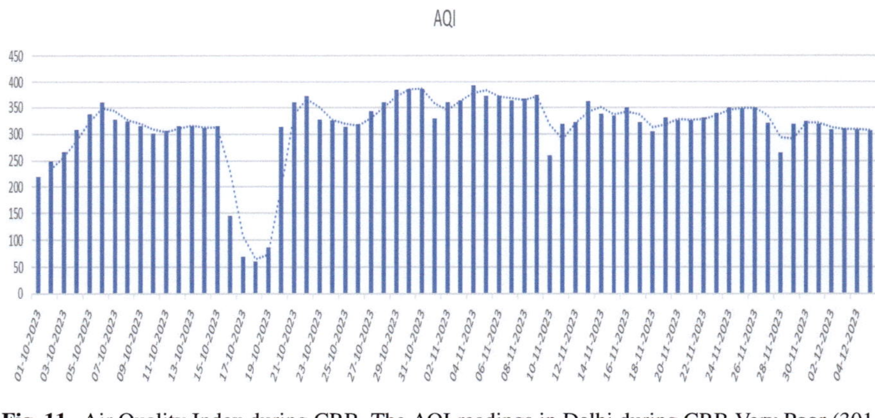

Fig. 11 Air Quality Index during CRB. The AQI readings in Delhi during CRB Very Poor (301–400) may cause respiratory illness to the people on prolonged exposure. Effect may be more pronounced in people with lung and heart diseases

4 Conclusion

These findings highlight the seasonal variability and the impact of factors such as agricultural crop burning, vehicular emissions, and weather conditions on air quality in Delhi. The substantial increases in PM_{10} and other pollutants during November, traditionally associated with crop burning in neighbouring states, underscore the need for comprehensive measures to mitigate air pollution and protect public health in the region.The significant impact of agricultural crop burning on air quality in Delhi during the winter months, highlighting the urgent need for effective measures to mitigate the adverse health effects associated with elevated levels of particulate matter and pollutants in the atmosphere [16].

Delhi's crop residue burning generates 17 times more particulate matter (PM) than all other sources combined, including car emissions, industry, and rubbish burning [17, 19]. Consequently, about 20 percent of the nation's organic and elemental carbon emissions from burning agricultural waste are produced by residue burning in northwest India [19]. Burning crops increase the amount of particles (PM) in the atmosphere, which has a significant impact on climate change. One of the factors contributing to global climate change is the release of primary and secondary fine black and brown carbon, which changes the absorption of light [14, 17, 18, 21]. $PM_{2.5}$ and PM_{10} particles in the air are commonly classified as fine and coarse particles, respectively, based on their aerodynamic diameter and chemical composition. Lightweight particles have a greater range of motion and a longer airborne suspension time [22]. The effect of particulate matter is intensified by meteorological circumstances because light particles stay in the air longer than heavy ones do. It was calculated that burning paddy residue contributed between 60 and 390 mg/m3 of $PM_{2.5}$ annually to the Patiala district of Punjab [20].

In conclusion, while crop residue burning primarily occurs in Punjab and Haryana, its impact on Delhi is profound due to atmospheric transport mechanisms and prevailing wind patterns. Addressing this issue requires concerted efforts at both regional and national levels to mitigate air pollution and its associated health and climate impacts.

In November and December, farmers in Punjab and Haryana often resort to crop burning as a traditional method to clear fields quickly for the next planting season. This practice, while efficient, contributes significantly to air pollution, impacting both local and regional air quality levels.

The Indian government, at the federal and state-level, has implemented strict rules and regulations outlawing burning practices, yet much of northern India still engages in the practice, especially Punjab, Haryana, and part of Uttar Pradesh. Patriotic adherence to these policies requires constant monitoring everywhere as well as prompt, efficient follow-up action.

Acknowledgements Authors would like to thank Director, CSIR-National Physical Laboratory, New Delhi, for providing the necessary support and infrastructure. The authors would like to express their sincere gratitude to Dr. C. Sharma (late) for his constant support. The authors would also like to thank Head of Air Lab CPCB, Govt. of India, and ENVEA PVT. LTD for their continuous support and encouragement. Authors also would like to thank the Academy of Scientific and Innovative Research (AcSIR) for facilitating as its Ph.D. scholar.

Declarations Funding: ENVEA Pvt. Ltd. India

Conflict of Interest: We have read, understood, and have complied, as applicable, with the statement on ethical responsibilities of authors. We declare that there are no conflicts of interest to disclose

Availability of Data: Data may be provided as per the Institutional guidelines

References

1. Singh D, Dhiman SK, Kumar V, Babu R, Shree K, Priyadarshani A, Saluja S (2022) Crop residue burning and its relationship between health, agriculture value addition, and regional finance. Atmosphere 13(9):1405
2. Kumar R, Mukherjee S, Debnath A, Mishra VN, Gupta K, Kumar K, Arya VS (2022) Climatology and spatio-temporal analysis of air pollution distribution in megacity of Delhi and its surrounding states. Available at SSRN 4053647
3. Kaushal LA, Prashar A (2021) Agricultural crop residue burning and its environmental impacts and potential causes–case of northwest India. J Environ Planning Manage 64(3):464–484
4. Sharma AR, Kharol SK, Badarinath KVS, Singh D (2010) Impact of agriculture crop residue burning on atmospheric aerosol loading–a study over Punjab State, India. In Annales Geophysicae. 28(2) pp 367–379. Copernicus GmbH
5. Basu M (2019) The great smog of Delhi. Lung India. 36(3):239–240. https://doi.org/10.4103/lungindia.lungindia_363_18. PMID: 31031347; PMCID: PMC6503731
6. Kedia S (2017) Delhi is Gasping for Air—Six reasons why the national capital is becoming a 'Gas Chamber'. (Last accessed on 2017 Nov 08)
7. Sarkar S, Singh RP, Chauhan A (2018) Crop residue burning in northern India: Increasing threat to Greater India. J Geophys Res: Atmos 123(13):6920–6934
8. Aladjadjiyan A (1992) Lessons from denmark and austria on the energy valorization of biomass (Contract No: JOU2-CT92–0212, Coordinator for Bulgaria). European Commission: Brussels, Belgium
9. Amonette J, Joseph S (2009) Characteristics of biochar: Micro-chemical properties. In: Lehmann J, Joseph S, editors. Biochar Environ Manag: Sci Technol. Earth Scan: London, UK. pp 33–52
10. Beck-Friis B, Pell M, Sonesson U, Jonsson H, Kirchmann H (2000) Formation and emission of N2O and CH4, from compost heaps of organic household waste. Environ Monit Assess 62:317
11. Franchi E, Agazzi G, Rolli E, Borin S, Marasco R, Chiaberge S et al (2016) Exploiting hydrocarbon-degrader indigenous bacteria for bioremediation and phytoremediation of amulti-contaminated soil. Chem Eng Technol 39:1676–1684
12. Gadde B, Bonnet S, Menke C, Garivait S (2000) Air pollutant emissions from rice straw open field burning in India. Thail Philippines Environ Pollut. 157:1554–1558
13. Hatch LE, Luo W, Pankow JF, Yokelson RJ, Stockwell CE, Barsanti KC (2015) Identification and quantification of gaseous organic compounds emitted from biomass burning using two-dimensional gas chromatography-time-of-flight mass spectrometry. Atmos Chem Phys 15:1865–1899
14. Mittal SK, Susheel K, Singh N, Agarwal R, Awasthi A, Gupta PK (2009) Ambient air quality during wheat and rice crop stubble burning episodes in Patiala. Atmos Environ 43:238–244
15. https://creams.iari.res.in/?page_id=1123
16. Hasan MR, Pattnaik S (2023) Characteristics of Western Disturbance intensification and associated induced circulations over the Indian region. Nat Hazards Res

17. Jain N, Bhatia A, Pathak H (2014) Emission of air pollutants from crop residue burning in India. Aerosol Air Qual Res 14(1):422–430
18. Rahman MH, Singh N, Kundu S, Datta A (2022) Potential areas of crop residue burning contributing to hazardous air pollution in Delhi during the post-monsoon season. 51(2), pp 181–192
19. Cusworth DH, Mickley LJ, Sulprizio MP, Liu T, Marlier ME, DeFries RS, Gupta P (2018) Quantifying the influence of agricultural fires in northwest India on urban air pollution in Delhi. India. Environ Res Lett 13(4):044018
20. Satyendra T, Singh RN, Shaishav S (2013) Emissions from crop/biomass residue burning risk to atmospheric quality. Int Res J Earth Sci 1(1):1–5
21. Lohan SK, Jat HS, Yadav AK, Sidhu HS, Jat ML, Choudhary M et al (2018) Burning issues of paddy residue management in north-west states of India. Renew Sustain Energy Rev 81:693–706
22. Saxena P, Sonwani S, Srivastava A, Jain M, Srivastava A, Bharti A, Bhardwaj S (2021) Impact of crop residue burning in Haryana on the air quality of Delhi, India. Heliyon, 7(5)

Analysis of Accuracy of Empirical and Semi-Empirical Models in Prediction of Sound Absorption Characteristics of Acoustical Materials

Anita Devi, Chitra Gautam, Naveen Garg, and Snehlata

Abstract The chapter reports the comparative evaluation of sound absorption characteristics predicted by various empirical and semi-empirical models such as Delany Bazley, Miki, Johnson-Champoux-Allard (JCA) and Johnson-Champoux-Allard-Lafarge (JCAL) models are used widely in building acoustics. These models are utilized in the present study for the prediction of sound absorption characteristics of commonly used acoustical materials. The parameters associated with various analytical models are presented in the present study and these models are then compared with the experimental study conducted in the Sound Impedance tube to check the accuracy.

Keywords Sound absorption · Impedance tube · Noise reduction coefficient · Empirical models

1 Introduction

Sound-absorptive materials absorb a large part of the sound waves incident on them and reflect a small part of the wave. These materials are very beneficial for sound control. One such effective sound-absorptive material is porous. To find the sound absorptive capability of a material we find its sound absorption coefficient, which is the ratio of sound intensity absorbed to the total incident sound intensity. The range of sound absorption coefficient values is 0 to 1, with a number closer to 1 indicating a more absorbent material.

A. Devi (✉) · C. Gautam · N. Garg · Snehlata
Academy of Scientific and Innovative Research (AcSIR), Uttar Pradesh, Ghaziabad, India
e-mail: anita.research2022@gmail.com

A. Devi
Government College Jatauli Haily Mandi, Gurugram, India

A. Devi · C. Gautam · N. Garg · Snehlata
CSIR-National Physical Laboratory, New Delhi, India

S. Yadav et al. (eds.), *Advances in Metrology*, Lecture Notes in Mechanical Engineering,
https://doi.org/10.1007/978-981-96-6418-4_15

When sound waves are incident on the porous material, air molecules inside and on the surface of the porous material vibrate with the incident wave which causes the molecules to lose some of their initial energy. This phenomenon occurs due to thermal and viscous losses at the walls of the material's internal pores and tunnels causing some of the energy in the air molecules to be transformed into heat [1]. These changes are adiabatic at high frequencies and isothermal at low frequencies [2]. Based on their microscopic structures, porous sound-absorbing materials can also be categorized as cellular, fibrous, or granular. Due to the increase in acoustical uses, there is an increase in the availability of absorptive acoustical materials. As lot of research is conducted on acoustical materials and major challenge is to predict the sound absorption characteristics as their testing cost is very high. Due to that instead of testing the absorptive behaviour of every material, there is a requirement for sound absorption prediction models. One can predict the acoustic performance of the acoustical material by knowing some of the non-acoustical parameters such as porosity, airflow resistivity, tortuosity, viscose, and thermal characteristic length [3]. Several studies conducted on theoretical models to predict acoustic behaviour and assess the effectiveness of porous materials [3–6]. The present study has an objective to evaluate and analyse the various empirical and microstructural/semi-empirical models developed so far for predicting the sound absorption coefficient values in the measurement frequency range of 100 Hz-5000 Hz.

2 Sound Absorption Prediction Models

Tang and Xiong [7] reviewed various empirical and microstructural sound absorption prediction models based on facile airflow resistivity and structural parameters of fibrous materials [7]. Zwikker and Kosten (1949) proposed the formula for the acoustic absorption coefficient for porous materials which consists of surface characteristic impedance, acoustic absorption, characteristic impedance, propagation constant, thickness, air density, and sound velocity [8]. The relationship between acoustic absorption and parameters is shown in the following Eq. (1) to (3) [8].

$$\alpha = 1 - \left| \frac{Z - W_0}{Z + W_0} \right|^2 \tag{1}$$

$$Z = Z_0 \coth \gamma l \tag{2}$$

and

$$W_0 = \rho_0 c_0 \tag{3}$$

where α is the sound absorption coefficient, Z is surface characteristic impedance, Z_0 is characteristic impedance, γ is propagation constant, l is the thickness, ρ_0 is the air density and C_0 is the sound velocity.

Delany and Bazley [4], using curve fitting, measured a huge number of samples of porous material and showed the relationships that explain how the propagation wavenumber and characteristic impedance change with flow resistivity [4]. Several researchers suggested in their review several types of models such as empirical models and microstructural models, to predict the sound absorption properties of the acoustical materials [3, 7].

2.1 *Macrostructural/ Empirical Models*

Equation (1) and (2) explain the empirical model to predict the absorption of porous absorbents by the non-acoustical properties of the material (flow resistivity, porosity, tortuosity and characteristic lengths) in terms of the characteristic impedance and (complex) wavenumber [9]. The theoretical models address how sound waves behave in the microstructural features of porous materials but porous materials have intricate geometrical and structural details, it is incredibly difficult to explain the acoustical behavior of many sound absorbers using theoretical models [10]. Consequently, empirical models with a macroscopic point of view are used to predict the sound absorption behavior of fibrous materials. This kind of empirical technique is the easiest to use and has the potential to be most successful when applied to a pre-existing material. Delany and Bazley's [4] model is the widely used empirical model for the prediction of sound absorption properties of acoustical materials [4]. Throughout a wide frequency range, they looked into the sound absorption characteristics of numerous fibrous materials with porosity close to unity [4]. By using Eq. (4) and (5), Delany and Bazley (1970) were able to determine the characteristic impedance, Z_C, and wave number, k_C [4]:

$$z_c = \rho_0 c_0 \left[1 + C_1 \left(\frac{\rho_0 f}{\sigma} \right)^{C_2} - iC_3 \left(\frac{\rho_0 f}{\sigma} \right)^{C_4} \right] \tag{4}$$

$$k_c = \frac{\omega}{c_0} \left[1 + C_5 \left(\frac{\rho_0 f}{\sigma} \right)^{C_6} - iC_7 \left(\frac{\rho_0 f}{\sigma} \right)^{C_8} \right] \tag{5}$$

where f = frequency, σ = flow resistivity of air, $\omega = 2\pi f$ = angular frequency, $C_1 = 0.0571$, $C_2 = -0.754$, $C_3 = 0.087$, $C_4 = -0.732$, $C_5 = 0.0978$, $C_6 = -0.7$, $C_7 = 0.189$, $C_8 = -0.595$ [4].

Thakur and Kessler (2016) suggested the boundary for the validity of the model in Eq. (6) [11]

$$0.01 < \left(\frac{\rho_0 f}{\sigma}\right) < 1.0 \quad (6)$$

In order to avoid giving negative values for the real part of the surface impedance, Miki [5] improved Delany and Bazley's [4] model. Dunne et al. [3] reviewed that the Miki model had superior prediction ability as compared to the original Delany and Bazley model, throughout the complete frequency range, particularly in the range $f < 400$ Hz [3] and [6] investigated the new model which can be used to predict the acoustical properties of a fibrous material with high accuracy compared with the Delany-Bazley and Miki models [6]. Some other empirical models such as Mechel [12], Garai Pompoli [13], Komatsu [6] etc. were developed with some adjustments to Eq. (4) and (5). One of the issues with these macroscopic empirical models is that they do not readily give information on the microscopic characteristics of the porous absorber such as pore size and orientation of the pores [9]. This indicates that beyond determining the necessary optimal flow resistivity, it is difficult to apply these empirical models to inform designs [9]. More intricate propagation models are needed to enhance designs.

2.2 *Microstructural/ Semi-Empirical Models*

The semi-phenomenological models treat the fibrous material frame as inflexible. The microscopic physical properties of acoustic wave propagation in fibrous materials serve as the foundation for microstructural approaches. The dynamic density and dynamic bulk modulus of these models are determined using four to eight non-acoustical factors (such as porosity, tortuosity, airflow resistivity, viscous and thermal characteristic lengths) [14]. The characteristic impedance and propagation constant can be determined by the equivalent bulk modulus $\tilde{K}_{eq}$ and dynamic bulk density $\tilde{\rho}_{eq}$ [14].

$$z_c = \left[\widetilde{\rho_{eq}}.\widetilde{K_{eq}}\right]^{1/2} \quad (7)$$

$$k_c = \omega\left[\widetilde{\rho_{eq}}/\widetilde{K_{eq}}\right]^{1/2} \quad (8)$$

Models Johnson-Champoux-Allard (JCA) [15] and Johnson-Champoux-Allard-Lafarge (JCAL) [15] are more frequently used as microstructural models. To describe porous materials with rigid or flexible frames, the JCA model needs five characteristics: porosity, flow resistivity, tortuosity, viscous characteristic length, and thermal characteristic length. These parameters define the material's dynamic density and bulk modulus [16]. In the JCAL model, static thermal permeability is introduced on the basis of the JCA model to establish the heat exchange model between solids and fluids at low frequencies, which contains six parameters [17]. To calculate dynamic density and bulk modulus, the Johnson-Champoux-Allard-Pride-Lafarge (JCAPL) model requires eight factors [18], making it more difficult to utilize.

3 Factors for Prediction of Sound Absorption

Characteristics Impedance and propagation constant are necessary to determine the sound absorption coefficient of any acoustical medium, as was covered in Eq. (1) and (2). The designers of absorbers need to know other factors to find characteristic impedance and propagation constant. There are many factors influencing the sound absorption characteristics of porous materials [3].

3.1 Airflow Resistivity

Airflow resistivity is a significant parameter that affects the acoustic characteristics of any material. It refers to the material's capacity to oppose the flow of air entering its core [19]. It measures the amount of air that can penetrate a porous material. The sound absorption characteristics of a material are influenced by the airflow resistance per unit depth of the material. While acoustic waves travel across an uneven and winding path, they create opposition, leading to a reduction in the amplitude of the wave and conversion of acoustic energy into heat [20]. If the value of flow resistivity is smaller, it indicates less opposition to flow across the media, while an increased value suggests that the pores are too close, impeding the flow of air, and hence, may result in lower sound absorption [21]. It becomes evident that as the flow resistivity increases up to a certain threshold, there is a notable rise in the sound absorption coefficient within the mid-frequency range[22]. Beyond this point, absorption decreases marginally and stabilizes. This phenomenon occurs due to the impedance mismatch between air and the absorbent, resulting in sound reflection from the front face, thereby diminishing the absorption of the porous material [22]. The sound absorption coefficient is predicted by practically all empirical models using this parameter.

3.2 Porosity

Porosity is defined as the proportion of the volume occupied by open pores to the total volume [23]. Open pores are specifically taken into account since they are the only ones that affect the absorption of incident sound [23]. The enhancement of porosity leads to an improvement in sound absorption capabilities [24]. The propagation of sound within the pores results in increased energy consumption, thereby enhancing the sound absorption performance [24]. For sound to be absorbed through friction, it must be able to enter the material, which requires an adequate number of surface pores [25]. When examining sound absorption in porous materials, it is crucial to consider the size of pores, the number of pores, and the type of pores [25]. Sagartzazu et al. [26] reviewed and suggested the porosity values for polymeric foams and fibre materials fall between 0.95 and 0.99 respectively [26]. Beyond the peak point, a rise in porosity results in a decline in sound absorption performance [24, 27]. Fouladi et al. [28] examined some natural fibres and observed that as porosity decreased, sound absorption increased. For the prediction of sound absorption, this parameter is necessary in various empirical and semi-empirical models, such as Miki, JCA, JCAL, JCAPL, etc.

3.3 *Tortuosity*

The absorption characteristics of a material are influenced by its tortuosity, which is a significant parameter. The term "tortuosity" was first introduced by Zwiker and Koster (1949) [7] to describe the interplay between the fluid and solid components of a permeable material and Egab et al. [29] reviewed and explained the tortuosity of materials are based on the fact that the propagation of sound through a saturated porous sample results in the production of secondary effects that can then be used to characterize the parameter [29]. It is a geometrical factor that represents the effect of the structure of internal pores on the fluid flow velocity through a porous material [30]. Tortuosity is also used to consider the uneven arrangement of pores in unit sections through the depth of the poro-elastic medium [26]. Figure 1 [31] depicts the sound propagation relation with tortuosity. As tortuosity increases, the sound absorption coefficient tends to rise towards the higher frequency range, However, in the low to intermediate frequency range (up to 1500 Hz), the absorption coefficient remains unaffected by tortuosity [23]. This parameter is one of the essential parameters for the prediction of sound absorption coefficient in JCA, JCAL and JCAPL models.

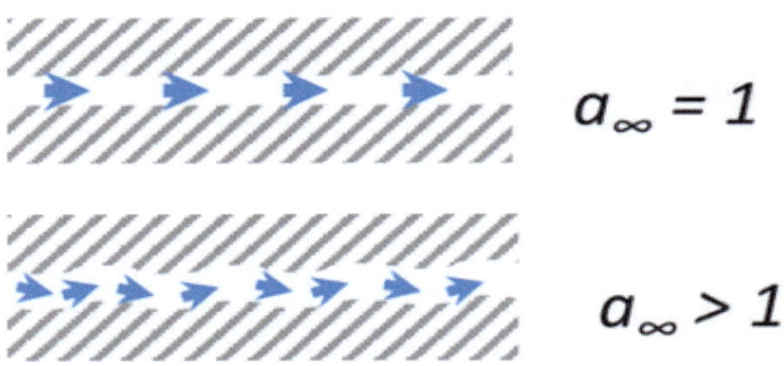

Fig. 1 Schematic representation of sound propagation in non-tortuous and tortuous paths [31]

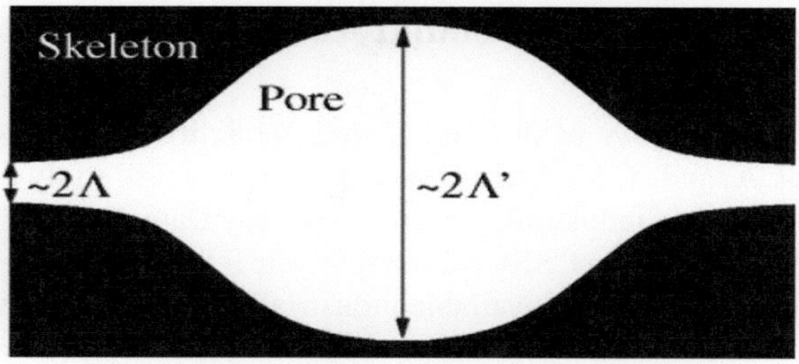

Fig. 2 Schematic figure depicting 2D pores with viscous characteristics length Λ and thermal characteristics length Λ′ [32]

3.4 *Viscous and Thermal Characteristics Length*

Viscous characteristic length (VCL) (Λ), serves to delineate the impact of viscosity on acoustical phenomena occurring at medium to high frequencies [32]. This parameter was originally proposed by Johnson et al. [15]. This parameter is linked to the extent of interconnection between two pores [32]. It denotes the average radius magnitude of the smaller cells and necks where viscous losses predominate over thermal losses [23]. Thermal characteristics length (TCL) (Λ′) is specifically twice the average ratio of volume to surface area of the pores [33, 34]. Figure 2 [32] depicts both the characteristic lengths. TCL was introduced by Champoux and Allard [34], building upon the groundwork laid by Johnson et al. [15] which is related to the thermal effect of porous materials at higher frequencies [22]. When the VCL is low, it exerts a greater influence on the SAC in the mid and high-frequency bands, resulting in lower sound absorption within those frequency ranges [22]. Reduced VCL leads to diminished fluid content within the porous sample, thereby decreasing its porosity and increasing tortuosity due to the rise in solid content within the material [22]. Both the characteristic lengths are used in JCA, JCAL, and JCAPL models.

3.5 *Static Thermal Permeability*

Just as viscous permeability plays a comparable role in the explanation of viscous forces, so too does static thermal permeability play in the description of thermal exchanges between frame and saturating fluid. Static thermal permeability is a geometrical parameter. When the thermal boundary layer is the same size as the characteristic size of the pores, this static thermal permeability partially describes the thermal effects at low frequencies. This parameter is used in JCAL and JCAPL models only for prediction of the sound absorption coefficient. Two additional factors, static thermal tortuosity and static viscous tortuosity, are necessary for the JCAPL model for the purpose of prediction [17]. Static thermal tortuosity is a similar parameter to low-frequency tortuosity, but instead of using the microscopic velocity field, it uses the microscopic temperature field. Static viscous tortuosity is another transport parameter related to viscous flow.

4 Accuracy of Sound Absorption Prediction Models

The accuracy of different sound prediction models, such as Delany Bazley, Miki, JCA, and JCAL, were assessed through a case study of several acoustical materials, including industrial-prepared coir and Date Palm Fiber. At the Sound Impedance tube facility in CSIR-National Physical Laboratory New Delhi, tests were conducted on commercially available industrially prepared coir with thicknesses of 20, 35, and 50 mm and densities of 99, 90, and 87 kg/m3 respectively, and date palm fiber with a thickness of 30 mm and density of 65 kg/m3 and a diameter of 30 mm. The algorithm proposed by these models was used to construct a Python code that is compatible with version 3.6.1 and analysed sound absorption properties. Gautam and Garg [35] discussed about the establishment of a Sound Impedance Tube facility at CSIR-NPL, New Delhi, which is used for acoustical materials testing [35]. Non-acoustical parameters of the acoustical materials used for the modeling are displayed in Table 1. In laboratory investigations, the Sound Impedance tube transfer function approach is employed because reverberation chamber experimentation can be laborious when working with samples covering greater areas [35]. Figure 3 shows the comparison of various sound absorption prediction models with Sound Impedance Tube experimental results. It is observed that empirical models i.e. Delany Bazley and Miki are showing more deviation from experimental results as compared to semi-empirical models (JCA and JCAL). JCA and JCAL models are showing similar results with JCAL having minimum deviation from experimental results comparatively. These prediction results are improving with an increase of the thickness of the sample. Noise reduction coefficient (NRC) which is the average of sound absorption coefficient values at four frequencies i.e. 250,500,1000 and 2000 Hz. NRC predicted value of the JCAL model is showing accurate results as compared to other sound absorption prediction models.

Table 1 Parameters utilized in the study

Material	Thickness (t) (mm)	Bulk density (kg/m^3)	Airflow resistivity (Pa.s/m^2)	Porosity (ϕ)	Tortuosity (α_∞)	VCL (Λ) (µm)	TCL (Λ') (µm)	References
Industrial prepared coir	20	99	1680	0.88	1.07	135	270	[36]
	35	90	1440	0.89	1.06	134	268	
	50	87	1359	0.89	1.06	133	266	
Date palm fiber	30	65	956	0.9278	2.9	247	430	[37]

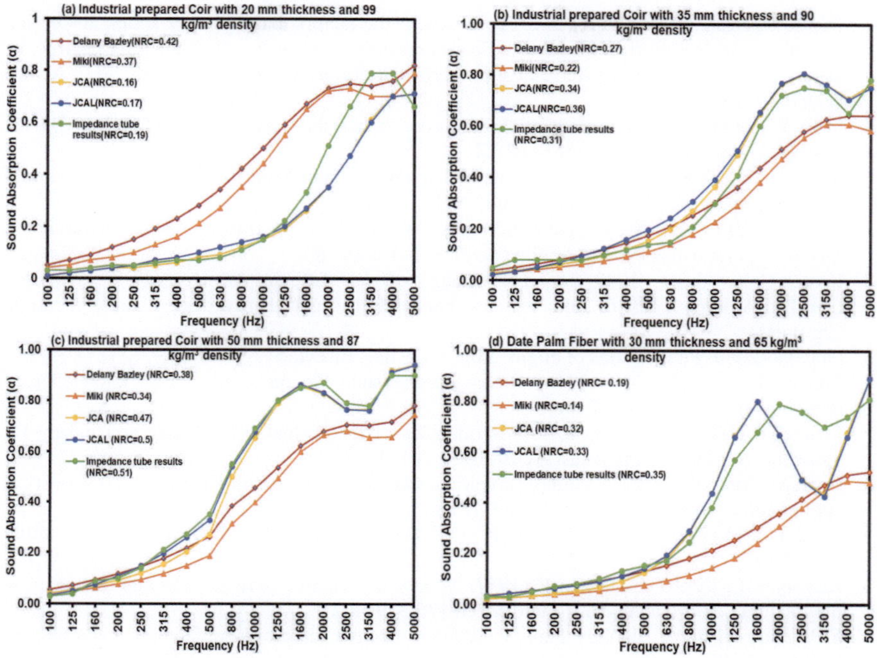

Fig. 3 Comparison of various empirical and semi-empirical models with impedance tube results

5 Conclusion

The non-acoustical parameters of the acoustical materials are used to investigate the empirical and semi-empirical models. The empirical models are quite simple as compared to semi-empirical models as they require a smaller number of parameters comparatively without going in detail of microstructure of the material. Empirical models are simple but could not provide the accuracy as provided by semi-empirical models. The JCA and JCAL models are showing similar results for acoustical samples. However, the JCAL model is more accurate showing less prediction error. With the increase of thickness of the sample, the prediction accuracy of both empirical and semi-empirical models is increasing.

References

1. Arenas JP, Crocker MJ (2010) Recent trends in porous sound-absorbing materials. Sound vibration 44(7):12–18
2. Allard J, Atalla N (2009) Propagation of sound in porous media: modelling sound absorbing materials. John Wiley & Sons

3. Dunne R, Desai D, Sadiku R (2017) A review of the factors that influence sound absorption and the available empirical models for fibrous materials. Acoustics Australia 45:453–469
4. Delany M, Bazley E (1970) Acoustical properties of fibrous absorbent materials. Appl Acoust 3(2):105–116
5. Miki Y (1990) Acoustical properties of porous materials-Modifications of Delany-Bazley models. J. Acoust. Soc. Jpn. 11(1):19–24
6. Komatsu T (2008) Improvement of the Delany-Bazley and Miki models for fibrous sound-absorbing materials. Acoust Sci Technol 29(2):121–129
7. Tang X (2017) Yan, Xiong "Acoustic energy absorption properties of fibrous materials: A review,." Composites Part A: Applied Science 101:360–380
8. Kosten C, Zwikker C (1949) Sound absorbing materials. Elsevier Amsterdam
9. Cox T, d'Antonio P (2016) Acoustic absorbers and diffusers: theory, design and application. Crc Press
10. Fahy FJ (2000) Foundations of engineering acoustics. Elsevier
11. Thakur VK, Kessler MR (2016) Green biorenewable biocomposites: from knowledge to industrial applications. CRC press
12. Mechel F (1976) Extension to low frequencies of the formulae of delany and bazley for absorbing materials. Acta Acust Acust 35(3):210–213
13. Garai M, Pompoli F (2005) A simple empirical model of polyester fibre materials for acoustical applications. Appl Acoust 66(12):1383–1398
14. Yang T et al (2020) Sound absorption properties of natural fibers: a review. Sustainability 12(20):8477
15. Johnson DL, Koplik J, Dashen R (1987) Theory of dynamic permeability and tortuosity in fluid-saturated porous media. J Fluid Mech 176:379–402
16. Yi W, Guo J, Zhou T, Jiang H, Fang Y (2024) A machine learning approach for predicting the Johnson-Champoux-Allard parameters of a fibrous porous material. Applied Acoustics 220, 109966
17. Jaouen L, Gourdon E, Glé P (2020) Estimation of all six parameters of Johnson-Champoux-Allard-Lafarge model for acoustical porous materials from impedance tube measurements. J Acoust Soc Am 148(4):1998–2005
18. Jaouen L (2024) JCAPL model. https://apmr.matelys.com/PropagationModels/MotionlessSkeleton/JohnsonChampouxAllardPrideLafargeModel.html
19. Hassan T et al (2021) Factors affecting acoustic properties of natural-fiber-based materials and composites: a review. Textiles 1(1):55–85
20. Hariprasad K, Ravichandran K, Jayaseelan V, Muthuramalingam T (2020) Acoustic and mechanical characterisation of polypropylene composites reinforced by natural fibres for automotive applications. J. Mater. Res. Technol. 9(6):14029–14035
21. Crocker MJ (2007) Handbook of noise and vibration control. John Wiley & Sons
22. Shravage P, Jain S, Karanth N (2010) Effect of intrinsic parameters on sound absorption and transmission loss-A parametric study. In INTER-NOISE and NOISE-CON congress and conference proceedings 2010, Institute of Noise Control Engineering 2010(2):664–673
23. Kalauni K, Pawar S (2019) A review on the taxonomy, factors associated with sound absorption and theoretical modeling of porous sound absorbing materials. J Porous Mater 26(6):1795–1819
24. Zhang Y, Li Y, Abdelhady A, Yang JJTRPDT (2020) Environment, Effect of different factors on sound absorption property of porous concrete, 87 p 102532
25. Seddeq HS (2009) Factors influencing acoustic performance of sound absorptive materials. Aust J Basic Appl Sci 3(4):4610–4617
26. Sagartzazu X, Hervella-Nieto L, Pagalday J (2008) Review in sound absorbing materials. Arch. Comput. Methods Eng. 15:311–342
27. Samsudin EM, Ismail LH, Kadir A, Nasidi IN (2017) Thickness, density and porosity relationship towards sound absorption performance of mixed palm oil fibers. In.: Proceedings of the 24th International congress on sound and vibration (ICSV 24), London, UK, pp 23–27
28. Fouladi MH et al (2013) Utilizing Malaysian natural fibers as sound absorber, Modeling measurement methods for acoustic waves for acoustic microdevices, pp 161–70

29. Egab L, Wang X, Fard M (2014) Acoustical characterisation of porous sound absorbing materials: a review. Int J Veh Noise Vib 10(1–2):129–149
30. Maderuelo-Sanz R, Barrigón Morillas JM, Martín-Castizo M, Gómez Escobar V, Rey Gozalo G (2013) Acoustical performance of porous absorber made from recycled rubber and polyurethane resin. Lat. Am. J. Solids Struct. 10:585–600
31. Acoustics A (2024) Alpha Acoustics. https://www.alfaacoustics.com/biot-parameters-characterization.html
32. APMR (2023). Acoustical porous material recipes, Propagation models assuming a motionless skeleton. https://apmr.matelys.com/
33. Leclaire P, Kelders L, Lauriks W, Melon M, Brown N, Castagnede BJJ (1996) Determination of the viscous and thermal characteristic lengths of plastic foams by ultrasonic measurements in helium and air 80(4):2009–2012
34. Champoux Y, Allard JFJJ (1991) Dynamic tortuosity and bulk modulus in air-saturated porous media 70(4):1975–1979
35. Gautam C, Garg N (2024) Establishment of sound impedance tube facility at CSIR-national physical laboratory, New Delhi, India. In: Recent Advances in Metrology, Singapore, pp 15–24: Springer Nature Singapore
36. Fouladi MH, Ayub M, Nor MJ (2011) Analysis of coir fiber acoustical characteristics 72(1):35–42
37. Taban E, Khavanin A, Jafari AJ, Faridan M, Tabrizi AK (2019) Experimental and mathematical survey of sound absorption performance of date palm fibers. Heliyon 5(6)

Study on Various Syringe Pumps and Infusion Device Analyzers for Low Liquid Flow Calibration Facility at CSIR-NPL, India

Sanjeet Kumar, Shiv Kumar Jaiswal, Anshul Varshney, Chatar Singh, Shanay Rab, and Sanjay Yadav

Abstract The present paper shows the study on the various syringe pumps and infusion device analyzers which are used in hospitals and research laboratories for precise drug dosing, especially for cancer patients receiving chemotherapy, insulin therapy, pain management and other critical applications. For the calibration of such instruments, the Fluid Flow Metrology Section of the CSIR-National Physical Laboratory (CSIR-NPL), New Delhi, has established a low-liquid flow calibration facility in the flow range of 1 to 1500 mL/h with an expanded uncertainty of 1.0 to 0.30% (at $k = 2$ for 95% confidence level). From the study of the calibration of the syringe pumps and infusion device analyzers, it is concluded that repeatability in flow measurement, evaporation and resolution of DUT are dominant factors of uncertainty at lower flow. However, for higher flow, the timer uncertainty and repeatability in measurement are dominant factors of uncertainty.

Keywords Syringe pump · Micro liquid flow · Accuracy · Uncertainty · Biomedical applications · Infusion device analyzer

1 Introduction

Low-liquid flow / micro-liquid flow measurement plays a crucial role in various industries and applications where precise control and monitoring of liquid flow rates are essential. Such applications include biotechnology, healthcare, sensor technology, process control, chemical analysis and synthesis, environmental monitoring, fuel

S. Kumar (✉) · S. K. Jaiswal · A. Varshney · C. Singh · S. Yadav
Fluid Flow Metrology Section, CSIR-National Physical Laboratory, New Delhi, India
e-mail: sanjeet11792@gmail.com

S. Kumar · S. K. Jaiswal · S. Yadav
Academy of Scientific and Innovative Research, Ghaziabad, India

S. Rab
School of Architecture, Technology and Engineering, University of Brighton, Brighton, UK

S. Yadav et al. (eds.), *Advances in Metrology*, Lecture Notes in Mechanical Engineering,
https://doi.org/10.1007/978-981-96-6418-4_16

cell technology, micro-reactors and lab-on-a-chip devices, semiconductor industry, aerospace and defence where the precise mixing, flow, or separation of different fluids or gases is essential, and maintaining a consistent volume or flow rate is crucial for optimal performance. For such applications, various types of low liquid flow meters such as microfluidic devices, mass flow meters, syringe pumps, infusion pumps and infusion device analyzers are used. The syringe pumps or infusion pumps are capable of maintaining a stable flow. Infusion tools are used in the clinical setting for patient feeding, hydration and medication administration. There are certain applications where the most critical criteria for optimal drug delivery are either total supplied volume (mass) or flow rate. For a major proportion of medicines, the precise flow rate is quite critical for optimal patient treatment. A flow rate uncertainty of 5% is enough for the most important medicines [1–3]. In India, the major applications of syringe pumps and infusion pumps are in hospitals, biomedical laboratories and R&D laboratories. There are two types of syringe pumps, (i) hospital grade syringe pumps and (ii) metrology grade syringe pumps. The hospital grade syringe pumps have usually low accuracy (±5% or better), however, the metrology grade syringe pumps have high accuracy (±1 % or better). The hospital grade syringe pumps are usually calibrated through infusion device analyzers as the reference standard having an accuracy of ±2% or better, thus maintaining a Test Uncertainty Ratio (TUR) better than 2.5:1. The infusion device analyzers are in turn calibrated through a high accuracy reference syringe pump or through a primary low-liquid flow standard. The accuracy of these reference syringe pumps is usually 0.50% or better in the flow range of 10 mL/h to 1500 mL/h. The primary standards for low-liquid flow are three types, i.e. (1) weighing type (2) interferometric type and (3) front tracking type. The flow is realized in the fundamental SI units of mass, length and time in all three primary standards. Keeping in view of the requirements of biomedical laboratories, hospitals, R&D Laboratories and other sectors, CSIR-NPL, India (i.e. the National Metrology Institute of India) has developed a micro (low) liquid flow calibration facility in the flow range of 1 to 1500 mL/h. The various syringe pumps and infusion device analyzers were calibrated through this primary low-liquid flow calibration standard. In the current paper, the calibration results of these syringe pumps and infusion device analyzers and conclusions drawn from these studies have been reported.

2 NPLI Low Liquid Flow Calibration Facility

The NPLI Low Liquid Flow Calibration Facility is based on the weighing method. It is comprised of a flow generation device, a device under Test (DUT), a precision balance equipped with a collection container, and a timer [4, 5]. Figure 1 shows the schematic diagram of this low-liquid flow calibration system.

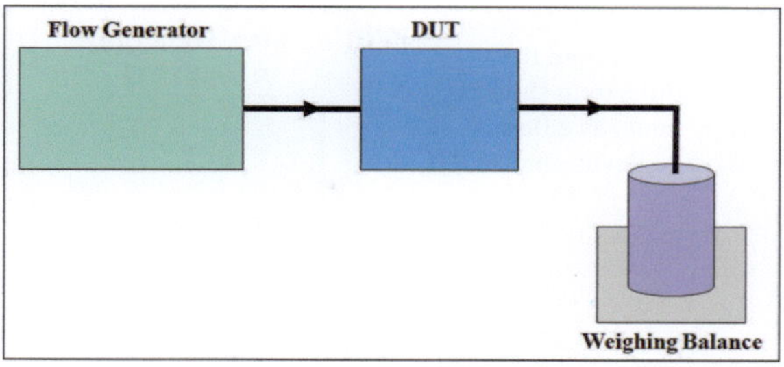

Fig. 1 Schematic diagram of NPLI low liquid flow calibration facility

2.1 *Flow Generation Device*

The syringe pump has been used as a flow generation device in the system. The syringe pump generates the required flow and it flows through DUT and finally to the precision balance. In the system, two syringe pumps have been used as a flow source to cover the flow rate from 1 to 1500 mL/h. The first syringe pump (Make: WPI, Model: SP100IZ, syringe volume: 10 mL) has the capability to provide flow from 0.1 to 125 mL/h whereas the second syringe pump (Make: Amogh Healthcare, Model: SP-1200, syringe volume: 50 mL) has the capability to provide flow from 0.1 to 1500 mL/h. The flow stability of the Amogh make syringe pump was found to be poor below 10 mL/h, therefore, it is used in the range of 10 to 1500 mL/h. The WPI make syringe pump is used for a lower flow range of 1 to 10 mL/h due to its better flow stability and limitation of syringe volume.

2.2 *Device Under Test (DUT)*

Different types of low liquid flow meters such as infusion device analyzers, syringe pumps, infusion pumps, mass flowmeters, etc. can be calibrated through this system. When the syringe pumps or infusion pumps are used as the DUT, then the schematic diagram shown in Fig. 1 changes to Fig. 2 where the syringe pump or infusion pump works both as a flow generation device and a DUT.

2.3 *Weighing Balance*

The 82 / 220 g (dual range) analytical balance (make: Radwag, Model: AS82/220.R2 PLUS) has been used in this system. The resolution of the balance is 0.00001 g in the 82 g range and 0.0001 g in the 220 g range. The glass beaker of 80 mL capacity has been used to collect the liquid and weighing it through balance. The balance is having the draft shield to protect the beaker from external influences such as air currents. The static weighing method has been used for flow measurement. A digital

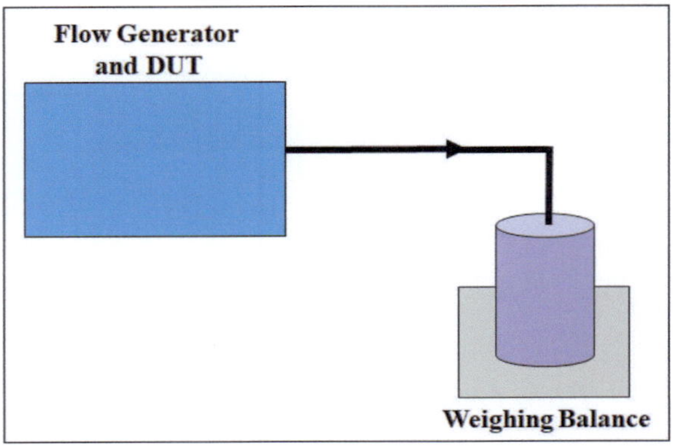

Fig. 2 Schematic diagram of NPLI low liquid flow calibration system showing syringe pump as a flow generation device and a DUT

timer of 0.01 s resolution has been used to calculate collection time which is used to derive the flow. The manual triggering is employed for the digital timer. The flexible PVC tubes are used to connect various instruments. The gauge no. 18 needle having an outer diameter (OD) of 1.27 mm and inner diameter (ID) of 0.838 mm is used at the end of the tube which connects it to the cork of the beaker. The cork has a hole of approx. 2 mm for smooth entry of the needle inside the beaker and it is placed on the top of the beaker. This reduces the evaporation effect. The temperature of the liquid in the beaker is measured using a high-precision digital thermometer, Make: Fluke, Model: 1523. This thermometer is having a resolution of 0.001 °C and an accuracy of ±0.02 °C.

3 Sources of Error and Uncertainty in Low Liquid Flow Measurement

The static-weighing method is used to measure four parameters i.e. (i) totalized mass (ii) mass flow rate (iii) totalized volume and (iv) volume flow rate. The various sources of errors and uncertainty in micro/ low liquid flow measurement are collected mass, collection time, liquid density, evaporation effect, flow rate stability, repeatability in measurement, zero drift of the DUT and resolution of the DUT. For a flow rate below 50 mL/h, a few drops of mineral oil are mixed with beaker water which forms a thin layer on the water surface, thus reducing the evaporation loss. The details about various sources of error and uncertainty have been described elsewhere [4–13], however, the equations for the derivation of mass flow rate, volume flow rate and uncertainty are reproduced here for the convenience of the readers.

The mass flow rate 'Q_m' is defined as the ratio of collected mass 'Δm' to collection time 'Δt' and it is represented by the following equation;

$$Q_m = \frac{\Delta m}{\Delta t} \quad (1)$$

The volume flow rate 'Q_v' is defined as the ratio of the mass flow rate to the liquid density 'ρ' and it is represented by the following equation;

$$Q_v = \frac{Q_m}{\rho} = \frac{\Delta m}{\Delta t.\rho} \tag{2}$$

The combined uncertainty for the volumetric flow rate is calculated using the following equation:

$$u_{Q_v} = \sqrt{u_{\Delta m}^2 + u_{\Delta t}^2 + u_{\rho}^2 + u_{EVAP}^2 + u_{REP}^2 + u_{RES}^2 + u_{ZD}^2 + u_{STAB}^2} \tag{3}$$

where

u_{Q_v} is the combined uncertainty of the volumetric flow rate.
$u_{\Delta m}$ is the standard uncertainty in the collected mass.
$u_{\Delta t}$ is the standard uncertainty in the collection time.
u_{ρ} is the standard uncertainty in density measurement of flowing liquid.
u_{EVAP} is the standard uncertainty due to evaporation,
u_{REP} is the repeatability in the error measurement of the DUT and.
u_{ZD} is the standard uncertainty due to zero drift of the DUT.
u_{STAB} is the standard uncertainty due to the stability of the water source.

The expanded uncertainty for the volumetric flow rate is determined by the following formula:

$$U_{Q_v} = k \times u_{Q_v} \tag{4}$$

where k is a coverage factor which is determined from the effective degree of freedom as per the Welch-Satterthwaite Formula given in the 'GUM' document [18].

4 Results and Discussion

In the present section, the calibration of syringe pumps and infusion device analyzers have been discussed. Some of the syringe pumps belong to customers who use the services of CSIR-NPL for their traceability. Such syringe pumps are (i) Infusion Pump, make: Harvard Apparatus, model: PHD2000, syringe volume: 50 mL, and (ii) Syringe Pump, make: Chemyx, model: Fusion 6000, syringe volume: 300 mL. The infusion device analyzers calibrated for the customers are of Fluke Biomedical make and having different models such as IDA-1S, IDA-4 and IDA-5. Before the use of a primary low-liquid flow calibration facility, the traceability of the system was established. The standards used in the facility i.e. analytical balance, digital timer, temperature and humidity indicator, barometer, digital thermometer and digital offline densitometer were calibrated through the national standards of their respective parameters [4, 5].

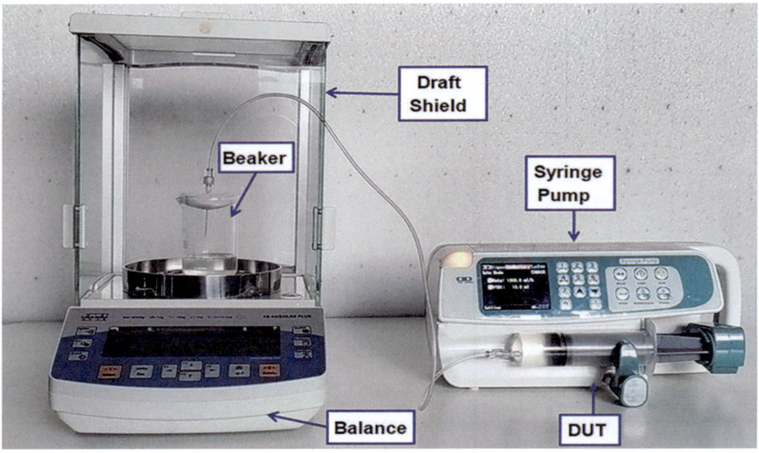

Fig. 3 Set-up for calibration of syringe pump, make: amogh, model: SP-1200

4.1 Calibration of Syringe Pump, Make: Amogh Healthcare

The syringe pump, make: Amogh Healthcare, model: SP-1200 was calibrated at various flow rates starting from 10 to 1500 mL/h using static weighing with the standing start and standing finish method. For its calibration, the schematic diagram shown in Fig. 2 was used. In this diagram, the syringe pump works both as a flow source and a DUT. Figure 3 shows the set-up for calibration of this syringe pump using low-liquid flow calibration facility. Tables 1 and 2 show the results of the calibration of the syringe pump in the totalized volume and volume flow rate mode at various flow rates. Since the syringe pump does not show the flow rate (at a particular set flow rate, it shows the dispensed/infused volume), therefore, for using it as a DUT for flow rate measurement, we considered a resolution of 0.01 mL/h for it because most of the flow measuring devices (e.g. infusion device analyzers) have a resolution of 0.01 mL/h.

4.2 Calibration of Syringe Pump, Make: Harvard Apparatus

The syringe pump, make: Harvard Apparatus, model: PHD2000 was calibrated at various flow rates starting from 10 to 1500 mL/h using a similar method as mentioned in section 4.1. For its calibration, the schematic diagram shown in Fig. 2 was used. Figure 4 shows the set-up for calibration of this syringe pump using low-liquid calibration facility. Tables 3 and 4 show the results of the calibration of the syringe pump in the totalized volume and volume flow rate mode at various flow rates.

Table 1 Results of calibration of the syringe pump, model: SP-1200, make: amogh healthcare in totalized volume mode

S. No	Nominal flow rate	Average DUT	Average standard collected volume	Relative error	Expanded uncertainty	Coverage factor k for approx.
	q	Collected volume V_{DUT}	V_{STD}	$(V_{DUT}-V_{STD}) \times 100/V_{STD}$	$U \pm$	95% CL
	(mL/h)	(mL)	(mL)	(%)	(%)	
1	10.0	10.000	10.285	−2.77	0.55	3.31
2	100.0	20.000	20.632	−3.06	0.29	3.31
3	400.0	50.000	51.823	−3.52	0.12	2.87
4	1000.0	50.000	51.712	−3.31	0.34	3.31
5	1500.0	50.000	51.741	−3.36	0.04	2.21

Table 2 Results of calibration of the syringe pump, model: SP-1200, make: amogh healthcare in volume flow rate mode

S. No	Nominal flow rate	Average DUT	Average standard flow rate	Relative error	Expanded uncertainty	Coverage factor k for approx.
	q	flow rate q_{VDUT}	q_{VSTD}	$(q_{VDUT}- q_{VSTD}) \times 100/q_{VSTD}$	$U \pm$	95% CL
	(mL/h)	(mL/h)	(mL/h)	(%)	(%)	
1	10.0	10.00	10.285	−2.77	0.55	3.31
2	100.0	99.91	103.070	−3.06	0.29	3.31
3	400.0	399.49	414.050	−3.52	0.12	2.65
4	1000.0	992.26	1026.222	−3.31	0.31	2.87
5	1500.0	1478.05	1529.514	−3.36	0.12	2.00

4.3 Calibration of Infusion Device Analyzer, Make: Fluke Biomedical

The infusion device analyzer, make: Fluke Biomedical, model: IDA-1S was calibrated at various flow rates starting from 10 to 1000 mL/h using the similar method as mentioned in section 4.1. For its calibration, the schematic diagram shown in Figure 1 was used where syringe pump has been used as flow source, although this can be directly calibrated using syringe pump, make: Harvard Apparatus, model: PHD2000 because the syringe pump model PHD2000 has the accuracy of 0.50% between the flow rate of 20 to 1500 mL/h whereas the infusion device analyzer model IDA-1S has the accuracy of 2% in this flow range, thus maintaining minimum test uncertainty ratio (TUR) of 3:1. Figure 4 shows the set-up for calibration of this

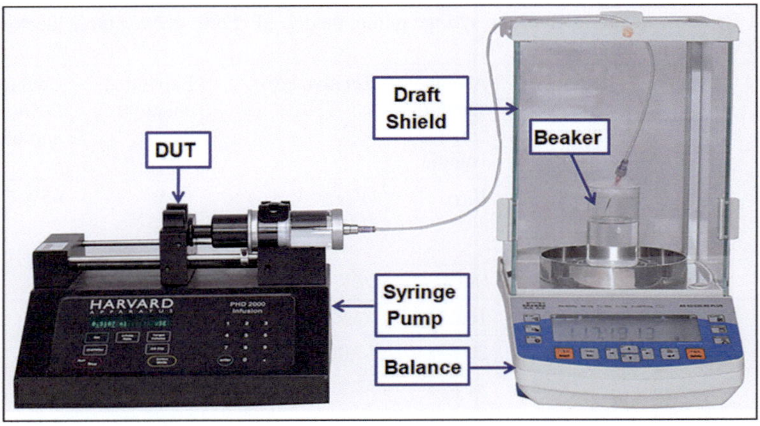

Fig. 4 Set-up for calibration of syringe pump, make: harvard apparatus, model: PHD2000

Table 3 Results of calibration of the syringe pump, model: PHD2000, make: harvard apparatus in totalized volume mode

S. No	Nominal flow rate	Average DUT	Average standard collected volume	Relative error	Expanded uncertainty	Coverage factor k for approx.
	q	collected volume V_{DUT}	V_{STD}	$(V_{DUT}-V_{STD}) \times 100/ V_{STD}$	$U \pm$	95% CL
	(mL/h)	(mL)	(mL)	(%)	(%)	
1	10.0	10.000	9.9390	0.61	0.53	3.31
2	100.0	20.000	19.9423	0.28	0.29	3.31
3	400.0	30.000	29.9295	0.24	0.11	2.87
4	1000.0	50.001	49.8751	0.25	0.24	3.31
5	1500.0	50.003	49.8842	0.24	0.05	2.23

infusion device analyzer using low-liquid flow calibration facility. Table 5 and 6 show the results of the calibration of the syringe pump in the totalized volume and volume flow rate mode at various flow rates (Fig. 5).

By observing the Tables 1, 2, 3 and 4 for syringe pumps, it is found that the expanded uncertainty for the totalized volume mode is in the range of 0.04 to 0.55% (for approx. 95% confidence level) for flow rates range of 10 to 1500 mL/h. Similarly, the expanded uncertainty for the volume flow rate mode is in the range of 0.11 to 0.55% (for approx 95% confidence level) for the same flow rate range. Observation of Tables 5 and 6 for infusion device analyzers shows this uncertainty in the range of 0.35 to 0.67%. Thus, the uncertainties at various flow rates do not exhibit a linear trend. Also, the observation of coverage factor k shows that at most of the cases (except in

Table 4 Results of calibration of the syringe pump, model: PHD2000, make: harvard apparatus in volume flow rate mode

S. No	Nominal flow rate	Average DUT	Average standard flow rate	Relative error	Expanded uncertainty	Coverage factor k for approx.
	q	flow rate q_{VDUT}	q_{VSTD}	$(q_{VDUT}-q_{VSTD}) \times 100/\, q_{VSTD}$ (%)	$U \pm$	95% CL
	(mL/h)	(mL/h)	(mL/h)	(%)	(%)	
1	10.0	10.000	9.9387	0.61	0.53	3.31
2	100.0	99.869	99.5855	0.28	0.29	3.31
3	400.0	399.674	398.7315	0.24	0.11	2.32
4	1000.0	998.107	995.6071	0.25	0.22	2.65
5	1500.0	1494.409	1490.9321	0.23	0.12	2.00

Table 5 Results of calibration of the infusion device analyzer, model: IDA-1S, make: fluke biomedical in totalized volume mode

S. No	Nominal flow rate	Average DUT	Average standard collected volume	Relative error	Expanded uncertainty	Coverage factor k for approx.
	q	collected volume V_{DUT}	V_{STD}	$(V_{DUT}- V_{STD}) \times 100/\, V_{STD}$	$U \pm$	95% CL
	(mL/h)	(mL)	(mL)	(%)	(%)	
1	20.0	10.02	10.1079	-0.84	0.54	2.87
2	100.0	30.13	30.4053	-0.92	0.35	2.87
3	600.0	51.38	51.8649	-0.94	0.46	2.87
4	1000.0	48.95	49.737	-1.58	0.52	2.87

Table 6 Results of calibration of the infusion device analyzer, model: IDA-1S, make: fluke biomedical in volume flow rate mode

S. No	Nominal flow rate	Average DUT	Average standard flow rate	Relative error	Expanded uncertainty	Coverage factor k for approx.
	(mL/h)	flow rate q_{VDUT}	q_{VSTD}	$(q_{VDUT}- q_{VSTD}) \times 100/\, q_{VSTD}$	$U \pm$	95% CL
		(mL/h)	(mL/h)	(%)	(%)	
1	20.0	20.01	20.163	–0.74	0.67	2.87
2	100.0	99.45	100.223	–0.77	0.45	2.37
3	600.0	595.09	600.947	–0.98	0.56	2.52
4	1000.0	994.21	1005.675	–1.14	0.54	2.17

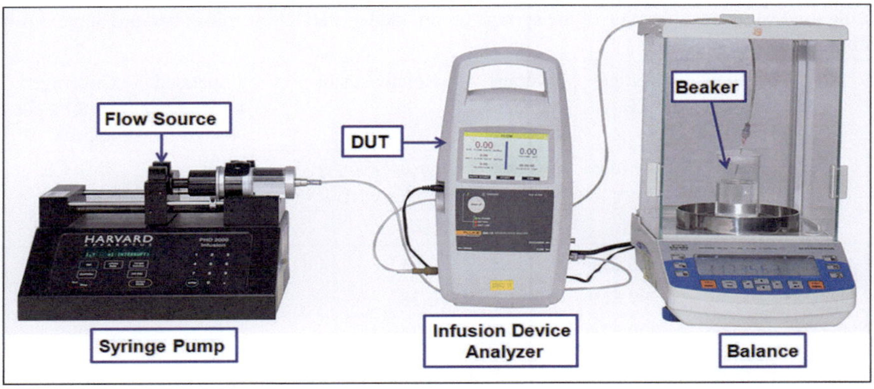

Fig. 5 Set-up for calibration of infusion device analyzer, make: fluke biomedical, model: IDA-1S

one case), the coverage factors are more than 2 leading to the conclusion that Type A uncertainty is higher than component Type B uncertainty. We also observe that Type A uncertainty is much higher at lower flow as compared to higher flow characterized by a higher coverage factor.

The results shown above do not represent the calibration and measurement capabilities (CMCs). For CMC purposes, the concept of the best existing device (BED) is applicable and we found that Chemyx syringe pump, model Fusion 6000 has the best repeatability and accuracy during the calibration [14]. Based on the results of this calibration, our claimed CMCs is 0.30% in the flow range 10 to 1500 mL/h. Similarly, for a lower flow range upto 1 mL/h, the WPI syringe pump, model: SP100IZ has been used. The results of the calibration of this syringe pump enabled us to claim a CMC of 1.0% down to 1 mL/h. These results have been shown elsewhere.

5 Conclusion

In the current paper, the various syringe pumps and infusion device analyzers were studied. The static weighing with the standing start and standing finish method was used for the flow measurement. These studies were carried out in the flow range of 10 to 1500 mL/h. From the studies, it is concluded that Type A uncertainty was the dominant factor of uncertainty in the lower flow rate due to poor repeatability in the measurement. The evaporation error was also the major source of uncertainty at lower flow which was reduced using an evaporation trap. At higher flow, the Type A uncertainty and timer uncertainty are dominant factors of uncertainty. This also leads to the conclusion that a syringe pump with lower syringe volume should be used at lower flow for better flow stability leading to better repeatability in the measurement and also for better flow accuracy. In the case of an infusion device analyzer, it is better to calibrate it using a reference syringe pump having an accuracy of ±0.50 %

or better in place of the gravimetric method as infusion device analyzers (IDAs) have an accuracy of ±2.0 % or better maintaining TUR of 3:1 or more. The calibration of IDAs is faster in the comparison method as compared to the weighing method as there is no requirement of minimum collection of mass in the weighing scale for better uncertainty or maximum collection time for better uncertainty of time measurement. The weighing method should be used for the best uncertainty.

Acknowledgements The authors express their gratitude to the Director, CSIR-National Physical Laboratory, New Delhi, for his unwavering encouragement and guidance. One of the authors, Sanjeet Kumar, extends his thanks to the Council of Scientific and Industrial Research (CSIR) in New Delhi, India, for the financial support in terms of fellowship.

References

1. Knight KH, Brand FM, Mchaourabiano AS, Veneziano G (2007) Implantable intracathal pumps for chronic pain; highlights and updates. Croat Med J 48(1):22–34
2. Hayek SM, Deer TR, Pope JE, Panchal SJ, Patel VB (2011) Intrathecal therapy for cancer and non-cancer pain. Pain Physician 14:219–248
3. Capes DF et al (1995) Fluctuations in syringe-pump infusions: association with blood pressure variations in infants. Am J Health-Syst Pharm 52(15):1646–1653
4. Kumar S, Jaiswal SK, Varshney A, Singh C, Rab S, Yadav S (2022) Uncertainty analysis of newly developed Low Liquid Flow Calibration Facility at CSIR-NPL", presented. In: The 11th International conference on advances in metrology (AdMet-2022), held at CSIR-National Physical Laboratory, New Delhi from August 24–26
5. Kumar S, Jaiswal SK, Varshney A, Singh C, Rab S, Yadav S (2024) range extension down to 1 ml/h of micro liquid flow calibration facility at CSIR-NPL. MAPAN-J Metrol Soc India (Under Press)
6. Batista E et al (2013) Calibration and use of syringe pumps. In: 16th International congress of metrology. EDP Sciences 02007
7. Batista E, Almeida N, Filipe E, Godinhol I (2014) Uncertainty calculation in gravimetric microflow measurements. St Petersburg: AMCTM 2014
8. Jaiswal SK, Yadav S, Bandyopadhyay AK, Agarwal R (2012) Global water flow measurement and calibration facilities: review of methods and instrumentations. MAPAN – J Metrol Soc India 27(2):63–76 (2012)
9. BIPM JCGM 200:2008, International vocabulary of metrology — Basic and general concepts and associated terms (VIM), 3rd Edition, 2012. Accessed on Aug 31 2024. http://www.bipm.org/utils/common/documents/jcgm/JCGM_200_2012.pdf
10. JCGM 100:2008 Evaluation of measurement data – Guide to the expression of uncertainty in measurement. Accessed on Aug 31 2024. https://www.bipm.org/documents/20126/2071204/JCGM_100_2008_E.pdf/cb0ef43f-baa5-11cf-3f85-4dcd86f77bd6
11. ISO 5168:2005, Measurement of liquid flow in closed conduits – Procedure for the evaluation of uncertainties, 2nd edition, pp 1–65, 2005
12. ISO 4185, Measurement of liquid flow in closed conduits – Weighing Methods, 1st edition, pp 1–21, 1980
13. Jaiswal SK, Singh C, Kumar S, Varshney A, Rab S, Yadav S (2022) Design and development for amelioration of primary water flow standard and calibration systems. Flow Meas Instrum 86 (102201): 1–12
14. WGFF Guidelines for CMC uncertainty and calibration uncertainty, October 21, 2013. Available in BIPM website at link. Accessed on Aug 31 2024. https://www.bipm.org/documents/20126/44756349/ccm-wgff-guidelines.pdf/eda74c50-1192-a20e-fc16-a3760e11329e

Comparison of Roundness Using Least Squares Reference Circle at Different Undulations Per Revolution

Sandeep Kumar, Jokhan Ram, and Mukesh Jewariya

Abstract Roundness of an object is crucial to determine for real world applications such as automobiles where many round objects are subjected to continuous motion. ISO 4291:1985 (ISO 4291:1985: Methods for the assessment of departure from roundness - Measurement of variations in radius) provides the method for evaluation of the deviation from the ideal roundness. This equals the difference of the maximum and the minimum radius of the measured profile of the object. We have assessed the out of roundness of a roundness artefact using a rotating spindle work table type instrument. The circle for assessing the out of roundness is the least square reference circle (LSCI) while using different undulations per revolution viz. 15 UPR, 50 UPR, 150 UPR and 500 UPR. A comparison of different UPRs is made and the effect of these UPRs on the roundness is summarised.

Keywords Roundness · Roundness standard · Dimensional metrology · Traceability · UPR · Filters

1 Introduction

Roundness is the deviation from a perfect circle and generally being measured to know the tolerance limit of any rotational part so that its functional aspects remain intact and does not lead to any failure or malfunction in the field. Various measurement techniques are available to measure the out of roundness of round objects namely diameter measurement method, Vee block, using coordinate measuring machine and rotational datum method (Fig. 1) [1]. Diameter measurement is perhaps the very method and easiest way to find the roundness and still in use when it comes to only aesthetics not the functional part. Vee block method has its own limitations and coordinate measuring machine generates small number of data which is not very much helpful when high accuracy is required. These limitations are covered by

S. Kumar (✉) · J. Ram · M. Jewariya
CSIR-National Physical Laboratory, New Delhi, India
e-mail: kumars2.nplindia@csir.res.in

S. Yadav et al. (eds.), *Advances in Metrology*, Lecture Notes in Mechanical Engineering,
https://doi.org/10.1007/978-981-96-6418-4_17

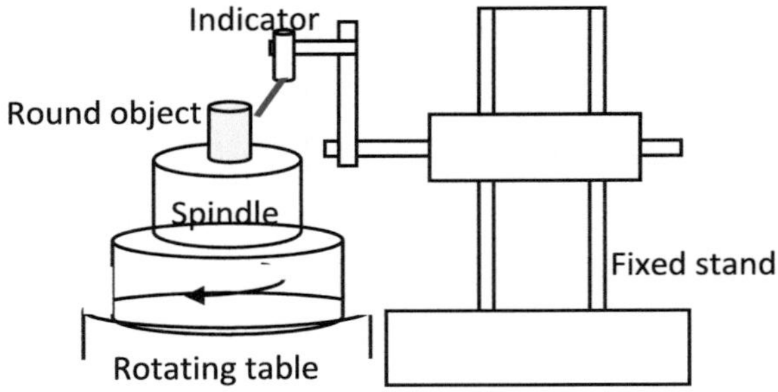

Fig. 1 Rotational datum method for roundness measurement

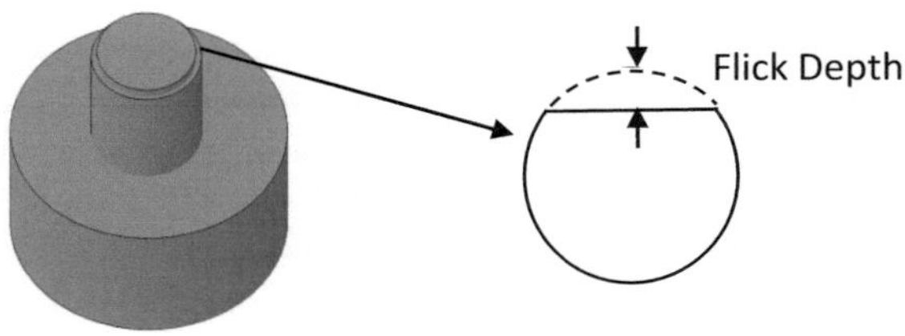

Fig. 2 Flick standard deviation

the rotational datum method and provides highly accurate results [1, 8]. At CSIR-National Physical Laboratory (CSIR-NPL) a rotational datum instrument is used to carry out the measurements [9].

2 Roundness Standards

Various roundness standards are out there which are used to confirm the values of any roundness measuring instrument in use. One such standard is flick standard. It is a round cylinder with a nominal depth flick (deviation) over its diameter. This deviation is demonstrated in Fig. 2.

3 Measurement Method

ISO 4291: 1985 recommends to assess the roundness departure by measuring the variation in radius. Roundness departure under reference conditions is expressed with reference to any of the following centers of:

- Least Squares Reference Circle (LSCI)
- Minimum Circumscribed Circle (MCCI)
- Minimum Zone Reference Circles (MZCI)
- The Maximum Inscribed Circle (MICI)

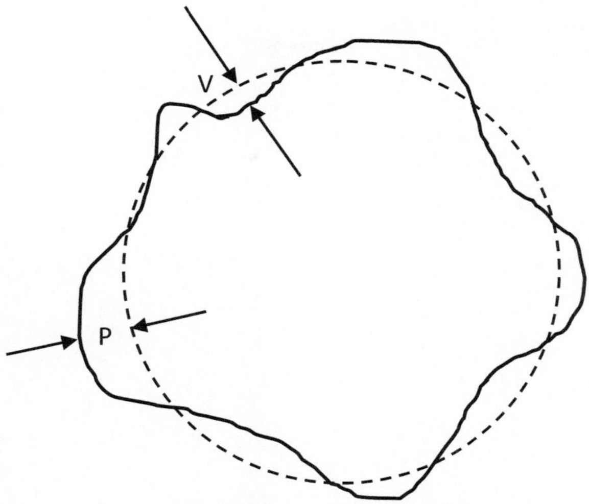

Fig. 3 Least squares reference circle (LSCI)

Least Squares Reference Circle is the most commonly used reference circle. LSCI is shown in Fig. 3.

LSCI measurement is performed using different undulations per revolution (UPR) along with Gaussian Filter.

4 Undulations Per Revolution

Undulations per revolution are the wavelength deviations on the surface from a perfect circle in one revolution and these deviations corresponds to form of the object [4]. During manufacturing, round objects are subjected to numerous processes which results in the different harmonics in terms of form deviations imprinted over the surface of the object. If these harmonics are not divided proportionally, then it is hard to understand the manufacturing information and its function. Hence UPR plays a critical role while ascertaining the out of roundness of any objects. These can be seen as wavelengths on the surface of the object but in a closed loop unlike roughness [5–7].

5 Results and Discussion

This part shows the measurement results of the flick standard and compared them with each other.

Results are compared with each other and shown in Table 1 and Fig. 4.

Table 1 RONt at different UPR

UPR	RONt, μm
15	6.161
50	13.652
150	16.190
500	16.543

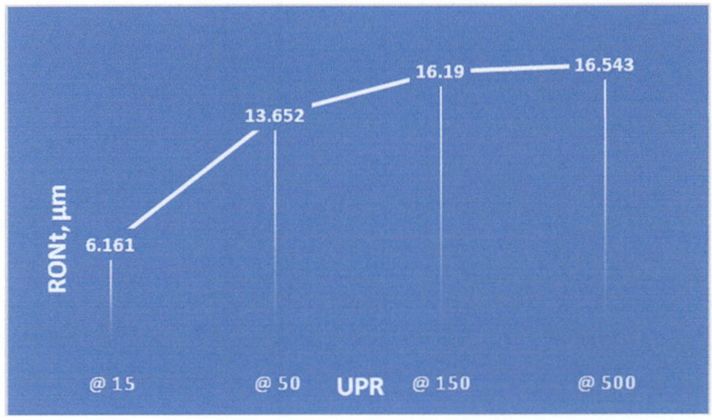

Fig. 4 Effect of different UPR on RONt

The measurement uncertainty [2] of RONt for the measurements is ± 0.030 μm @ $k = 2$.

6 Conclusion

This paper shows the out of roundness measurement results of a flick standard using LSCI reference circle. Separate undulations viz 15 UPR, 50 UPR, 150 UPR and 500 UPR are used during the measurement and these results are compared with each other.

The main observation is that at lower undulations, we lose significant information about the roundness deviations which may impact the function of objects while in the field. In our case, flick standard exhibits maximum information of roundness deviation while subjected to 150 UPR minimum during measurements. At higher spectral number defining undulations per revolution, we see that appropriate information is obtained and this information can be used to ascertain the manufacturing quality as well as the function of the object while in action. It is advisable to use higher undulations per revolution to extract maximum information about the out of roundness.

References

1. Taylor Hobson, What is roundness | methods of roundness measurement in metrology. www.taylor-hobson.com
2. ISO Guide to the expression of uncertainty in measurement
3. Whitehouse DJ (1967) An improved type of wave filter for use in surface finish measurement, proceedings institution of mechanical engineers, Part B. J Eng Manuf 192:306–318
4. ISO 4291:1985: Methods for the assessment of departure from roundness - Measurement of variations in radius
5. ISO 11562, Surface texture: profile method metrological characteristics of phase correct filters, 1996
6. Undulations per revolution (UPR) data in the roundness graphs. https://www2.mitutoyo.co.jp
7. Whitehouse DJ (2002) Surfaces and their measurement
8. Thalmann R (2006), Basic of highest accuracy roundness Measurement, Simposio de Metrología
9. Jewariya M, Ram J, Moona G, Singh G, Singh A, Kumar S, Sharma R (2022) Calibration of glass hemisphere and traceability at CSIR-National Physical Laboratory-India, MAPAN. https://doi.org/10.1007/s12647-022-00562-5

High-Precision Refractive Index Measurement of a Hydrophilic Sample Using Laser Interferometry

Kuldeep Kumar, Vikas, Surya Kumar Gautam, Mukesh Jewariya, and Manoj Das

Abstract This research paper presents an in-depth investigation into the application of laser interferometry for the precise measurement of refractive indices in diverse materials. Leveraging the principles of interference, the study focuses on developing a robust methodology that offers accurate and reliable refractive index data. The experimental setup employs laser interference techniques, specifically utilizing a Mach–Zehnder interferometer, to achieve high-resolution measurements. The interference patterns obtained, comprising alternating bright and dark fringes, are analysed to determine the refractive index of the material under investigation. The findings contribute to advancing the understanding of laser interferometry as a powerful tool for refractive index characterization, offering valuable insights for both theoretical developments and practical applications.

1 Introduction

Refractive index measurement is a critical technique extensively utilized in scientific and industrial domains to analyze the optical properties of various materials [1, 7]. Defined as the ratio of the speed of light in a vacuum to its speed in a medium, the refractive index offers valuable insights into the composition, concentration, and purity of substances [2]. Accurate refractive index data are indispensable for applications such as pharmaceutical quality control, biochemical solution analysis, optical material design, and environmental monitoring. The need for precise refractive index measurement methods has grown considerably, especially for complex sample matrices like water-soluble compounds. These samples, including pharmaceutical formulations, biochemical solutions, and aqueous suspensions, present unique challenges due to their sensitivity to environmental factors such as temperature, pH, and

K. Kumar (✉) · Vikas
Academy of Scientific and Innovative Research (AcSIR), Ghaziabad, India
e-mail: kuldeep20051995@gmail.com

K. Kumar · Vikas · S. K. Gautam · M. Jewariya · M. Das
CSIR-National Physical Laboratory (CSIR-NPL), New Delhi, India

S. Yadav et al. (eds.), *Advances in Metrology*, Lecture Notes in Mechanical Engineering,
https://doi.org/10.1007/978-981-96-6418-4_18

ionic strength, alongside their variable compositions. Ensuring accurate characterization of these samples is essential for maintaining the consistency and efficacy of final products, particularly in the pharmaceutical and biotechnology sectors.

Optical interferometric techniques fundamentally rely on interference patterns formed by the superposition of light waves from coherent sources traveling in different directions [2, 7]. These patterns, consisting of alternating bright and dark fringes, reflect local phase shifts caused by differences in the optical paths of the interfering beams [2]. Constructive interference, where waves are in phase, produces bright fringes, corresponding to an optical path difference equal to an integral multiple of the wavelength. Conversely, destructive interference occurs when waves are out of phase, creating dark fringes with an optical path difference equal to an odd multiple of half the wavelength. Interferometric methods are categorized based on how coherent beams are generated from a single light source [3, 4]. The first category, known as wavefront division, divides the wavefront from a common source (e.g., a point or narrow slit) using mirrors, biprisms, or lenses. Instruments such as the Fresnel biprism, Fresnel mirrors, and Lloyd's mirror exemplify this approach [3, 4]. The second category, amplitude division, splits the amplitude of the incident beam through partial reflection or refraction to produce coherent beams traveling in different directions, which later recombine to form interference patterns. Examples include Michelson's interferometer, Fabry–Perot interferometer, and Mach–Zehnder interferometer [3, 5, 6].

Interference pattern variations provide critical information about the medium through which one of the beams passes. For instance, fringe pattern variations have been utilized to determine gas parameters in cylindrical discharge tubes and to map the radial and axial distribution of a gas's refractive index in real-time [8]. The rotational temperature of the OH free radical was derived by analyzing phase shifts caused by the refractive index of the studied medium [9]. Shevchenko et al. [10] employed an improved Mach–Zehnder interferometer to calculate evaporation rates and vapor-in-air diffusion coefficients by monitoring refractive index changes. Their work highlighted digital interferometry's capability for measuring optical and diffusion properties of vapor-air mixtures, presenting precise data for acetone and HFE-7100 vapors at ambient temperatures ranging from 283.15 to 313.15 K [10]. Boudaoud and Lemerini [10] used a Mach–Zehnder interferometer to diagnose and visualize spatial density maps of pure nitrogen gas under a stationary corona discharge at pressures of 220–760 Torr. Their findings revealed inhomogeneous nitrogen density distribution due to pressure variations and linked gas density evolution to electric field changes influenced by space charge [10].

Our experimental methodology integrates a Mach–Zehnder interferometer with meticulous sample preparation to deliver reliable and reproducible refractive index measurements. Renowned for its high sensitivity and precision, the Mach–Zehnder interferometer detects minute changes in optical path lengths corresponding to refractive index variations in samples. By systematically controlling environmental conditions and varying sample concentrations in aqueous solutions, we validate the effectiveness of our approach in accurately characterizing the refractive indices of water-soluble compounds. This method not only enhances measurement accuracy but also provides deeper insights into the behavior of complex aqueous systems, supporting quality control, research, and development efforts across diverse industries [1].

2 Principle

The Mach–Zehnder interferometer splits a coherent laser beam into two paths using a beam splitter BS1. The object beam reflected from the mirror M1 passes through the sample, while the other reflected from the mirror M2 serves as a reference. The two beams are then recombined at another beam splitter BS2, creating an interference pattern. The setup is shown in Fig. 1. The change in the interference pattern by varying the concentration of salt in the water is observed and this change is used to calculate the change in refractive index of the water. The object and reference beams are tilted slightly to produce off-axis interference pattern. The object field from the recorded interference pattern can be reconstructed in a single shot. The reconstruction technique is as follow,

Interference Pattern contains the superposition of two fields and can be written mathematically as

$$I(x, y) = |E_0(x, y) + E_r(x, y)|^2 \tag{1}$$

Fig. 1 The mach–zehnder interferometer setup

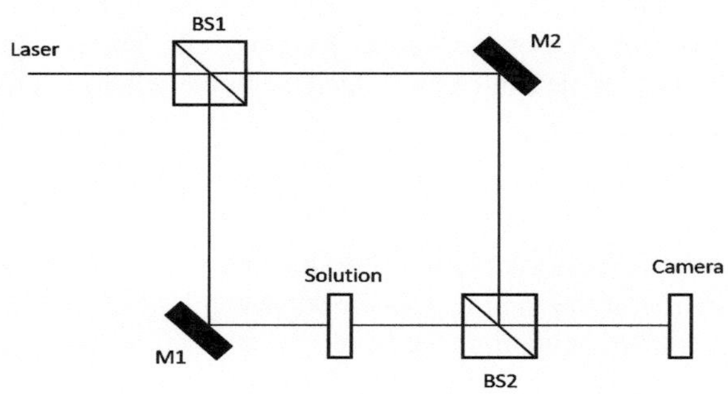

where $I(x, y)$ is the intensity of the interference pattern, $E_0(x, y)$ and $E_r(x, y)$ represent the object and reference fields. By expanding the intensity expression, Eq. (1) can be modified to

$$I(x, y) = |E_0(x, y)|^2 + |E_r(x, y)|^2 + E_0(x, y)E_r^*(x, y) + E_0^*(x, y)E_r(x, y) \tag{2}$$

Here the terms can be interpreted as $|E_0(x, y)|^2$ intensity of the object beam, $|E_r(x, y)|^2$ intensity of the reference beam, $E_0(x, y)E_r^*(x, y)$ the interference term, $E_0^*(x, y)E_r(x, y)$ the conjugate interference term. By tilting the reference beam, overlapping between the interference $E_0(x, y)E_r^*(x, y)$, and $E_0^*(x, y)E_r(x, y)$ conjgate interference terms can be avoided. These terms are spatially separated in the Fourier domain. The Fourier transform f of the interference pattern can be written as,

$$\begin{aligned} f\{I(x, y)\} =& f\left\{|E_0(x, y)|^2\right\} + f\left\{|E_r(x, y)|^2\right\} + \\ & f\left\{E_0(x, y)E_r^*(x, y)\right\} + f\left\{E_0^*(x, y)E_r(x, y)\right\} \end{aligned} \tag{3}$$

Since the 3rd term $f\left\{E_0(x, y)E_r^*(x, y)\right\}$ is separated from the 4th term $f\left\{E_0^*(x, y)E_r(x, y)\right\}$ due to the off-axis configuration, the 3rd term can be cropped in Fourier domain. By operating inverse Fourier transformation operation f^{-1} on this cropped term, one can obtain the 3rd term.

$$E_0(x, y)E_r^*(x, y) = f^{-1}\left\{f\left\{E_0(x, y)E_r^*(x, y)\right\}\right\} \tag{4}$$

The phase $\phi(x, y)$ can be directly obtained using the relation,

$$\phi(x, y) = \arg\left(E_0(x, y)E_r^*(x, y)\right) \tag{5}$$

The phase difference can be measure by recording another pattern by varying the concentration of the water by adding salt. The change in refractive index n of the liquid sample can be calculated using the relation as mentioned below

$$n = \frac{\Delta\varphi \cdot \lambda}{2\pi . d} \tag{6}$$

where $\Delta\varphi$ is the phase difference observed before and after changing the concentration of the sample, λ is the wavelength of the light used, d is the thickness of the qubit/cell containing the water.

Component-Wise Description

- Coherent Light Source (e.g., Laser) wavelength 632.8 nm, and a power of 2 mW.
- 2 Beam Splitters (50% transmission and 50% reflection), BS1 and BS2.
- 2 Mirrors, M1 and M2.
- Qubit/sample cell is a transparent container that holds the liquid solution.
- 8-bit CMOS Camera for detecting the interference pattern.

In designing the setup, one needs to take some precautions before taking the data, ensure the interferometer is stable and free from vibrations during measurements. Use clean and transparent sample cells to avoid scattering or absorption of light. Handle the coherent light source carefully to avoid damage or misalignment. Ensuring that all optical components (beam splitters, mirrors, and detector) are properly aligned. Adjust the mirrors so that the two light beams travel identical optical paths before recombining. Place the sample cell containing the water solution in one of the paths of the interferometer. Ensure that the cell is positioned such that the light beam passes through the water without any deviation.

3 Experimental Result

Interference patterns recorded before and after changing the concentration of the water are shown in Fig. 2a, b. Change in refractive index can be measured by using the Eq. (6) and from the phase difference information as shown in Fig. 2c.

4 Conclusion

The change in refractive index of the water is determined by analysing and comparing the interference patterns captured before and after adding the salt using the Mach–Zehnder interferometer. The precision of the measurement depends on the stability of the setup, the quality of the optical components, and the environmental conditions.

Fig. 2 Experimental result

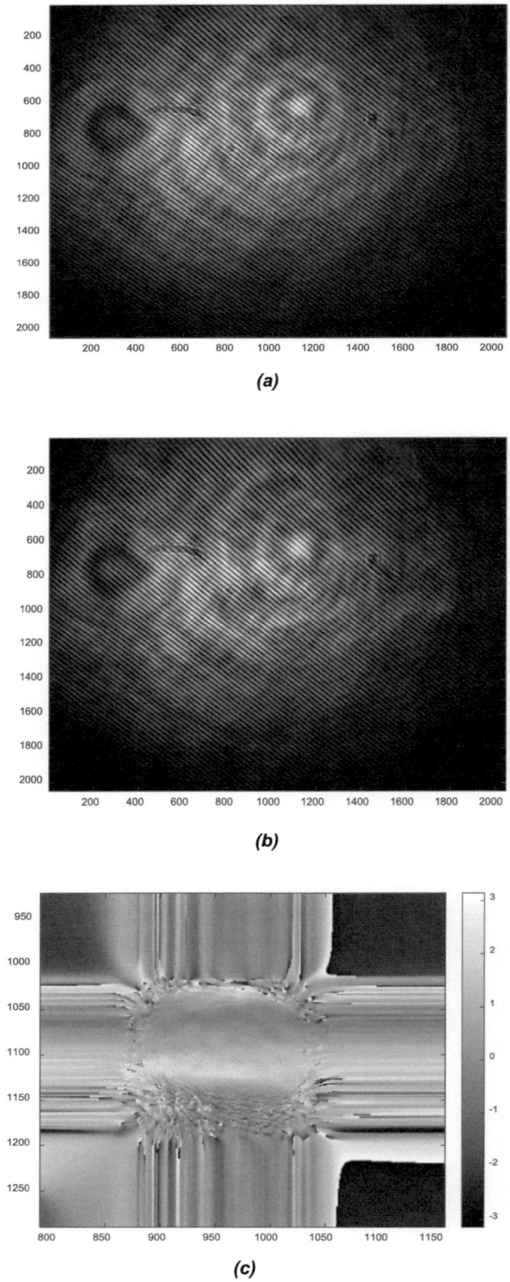

References

1. Wang J, Wang DN, Ge Y, Xu B, Guo X (2022) Optical fiber mach zehnder interferometer based on a pair of microholes. Opt Fiber Technol 73:103016
2. Boudaoud F, Lemerini M (2015) Using a mach-zehnder interferometer to deduce nitrogen density mapping. Chin Phys B 24:075205
3. Stafe M, Marcu A, Puscas NN (2014) Pulsed laser ablation of solids. Springer Series in Surface Sciences. (Springer, Berlin, Heidelberg, 2014) p. 53
4. Hariharan P (2003) Optical Interferometry, 2nd edn. Elsevier, Amsertedam
5. Ghazy AR, Shaheen M, Kenawy E, El-Mekawy F (2016) Determination of chitosan scattering parameters by using multi angle static laser light scattering (MASLLS). Delta J Sci 37:69–75
6. Shaheen ME, Ghazy AR, Kenawy E, Ghazy RA, ElMekawy F (2020) Experimental studies on static laser light scattering of synthesized poly (acrylonitrile-co- methyl methacrylate) copolymer at room temperature. Optik (Stuttg) 224, 165773
7. Shaheen ME, Abdelhameed ST, Abdelmoniem NM, Hashim HM, Ghazy RA, Abdel Gawad SA, Ghazy AR (2024) Determination of the refractive index of air and its variation with temperature and pressure using a Mach–Zehnder interferometer
8. Rosui N, Ralea M, Popescu IM, Iova I, Paraschiv R, Mircea D (1994) A holographic method for determination of the refraction index of a gas in electric discharge tubes. Pure Appl. Opt. J. Eur. Opt.Soc. Part A 3:859–865
9. Rabat H, De Izarra C (2004) Check of OH rotational temperature using an interferometric method. J Phys D Appl Phys 37:2371–2375
10. Shevchenko V, Mialdun A, Yasnou V, Lyulin YV, Ouerdane H, Shevtsova V (2021) Investigation of diffusive and optical properties of vapour-air mixtures: The benefits of interferometry. Chem Eng Sci 233:116433
11. Hendawy N, Said S, Nafie H, Mekawey FE (2008) Precise determination of refractive index of most popular environmental pollutant gases (1). Egypt J Phys 39:11–24
12. Prasad PR, Selvaraja SK, Varma M (2018) Real-time compensation of errors in refractive index shift measurements of microring sensors using thermo-optic coefficients. Opt Express 26:13461

Frequency Characterization of AC Resistance Standards Using Electrical Equivalent Circuit Model

Anshul, Somya Dixit, Aditi Sikarwar, Priyanka Jain, Suraj Khanna, and Satish

Abstract The frequency characteristic of AC resistance standards is essential for metrological applications as it bridges the gap from audio frequencies to radio frequencies in impedance metrology. In this proposed work, the traceability of AC resistance standards has been established using an electrical equivalent circuit model (EECM) upto 1 MHz. The AC resistance is measured using a precision LCR meter, and it is compared with the theoretical values obtained from the EECM at specified frequencies. The uncertainty in measurement for AC resistance standards has been computed according to the Guide to the Expression of Uncertainty in Measurement (GUM) document.

1 Introduction

Frequency characteristics of ac resistance are crucial for many applications and the change in the ac resistance standards occurs due to their parasitic and residual components. In this work, the realization of ac resistance is proposed using electrical equivalent circuits model (EECM) [1, 2]. The impedance may be expressed as [3]:

$$Z = R_f + jX \tag{1}$$

Here, R_f represents the real part, also known as AC resistance, while X denotes the imaginary part, also referred to as reactance. The electrical equivalent circuit model of an AC resistance is shown in Fig. 1. In the proposed work, a four-terminal pair AC resistance set (Agilent Model: 16074A) is used for the frequency characterization upto 1 MHz.

Anshul · S. Dixit · A. Sikarwar · P. Jain · S. Khanna · Satish (✉)
Electrical and Electronics Metrology Division, CSIR-National Physical Laboratory, New Delhi, India
e-mail: satish.nplindia@csir.res.in

Anshul · S. Dixit · A. Sikarwar
Banasthali Vidyapith, Newai, Rajasthan, India

S. Yadav et al. (eds.), *Advances in Metrology*, Lecture Notes in Mechanical Engineering,
https://doi.org/10.1007/978-981-96-6418-4_19

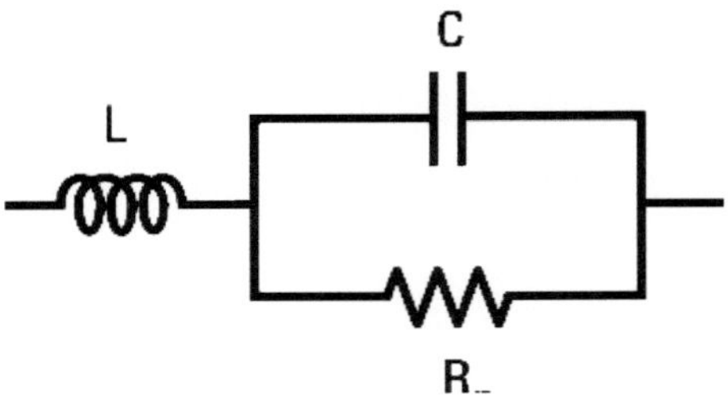

Fig. 1 Electrical equivalent circuit model of resistance

2 High-Value AC Resistance

For high value resistance, the parallel RC network will dominate as shown in Fig. 1. Thus, AC resistance may be given by:

$$R_f = \frac{R}{1 + (\omega RC)2} \tag{2}$$

Here, ω represents the angular frequency, R is the reference resistance at 1 kHz, and C denotes the parasitic capacitance.

3 Low-Value AC Resistance

For low-value resistance standards, Series R and L network will dominate and the electrical equivalent circuit simplifies to a parallel combination of conductance and susceptance. Thus, AC resistance of a low-value resistance standard may be given by:

$$R_f = R\left(1 + \left(\frac{\omega L}{R}\right)^2\right) \tag{3}$$

4 Uncertainty in Measurement

I. **High Value Resistance**

$$\Delta R_f = \sqrt{\left(\left(\frac{\partial R_f}{\partial R}\Delta R\right)^2 + \left(\frac{\partial Rf}{\partial C}\Delta C\right)^2\right)} = \sqrt{(c_1 \Delta R)^2 + (c_2 \Delta c)^2} \tag{4}$$

where

$$c_1 = \frac{1}{1 + (\omega CR)^2} - \frac{2(\omega CR)^2}{\left[1 + (\omega CR)^2\right]^2}$$

$$c_2 = \frac{-2\omega^2 R^3 C}{\left[1 + (\omega CR)^2\right]^2}$$

II. **Low Value Resistance**

$$\Delta R_f = \sqrt{\left(\left(\frac{\partial R_f}{\partial R}\Delta R\right)^2 + \left(\frac{\partial Rf}{\partial C}\Delta L\right)^2\right)} = \sqrt{(c'_1 \Delta R)^2 + (c'_2 \Delta L)^2} \quad (5)$$

where

$$c'_1 = -\left(\frac{\omega L}{R}\right)^2 + 1$$

$$c'_2 = \frac{2\omega^2 l}{R}$$

5 Measurement Setup and Observations

The precision LCR meter (Keysight E4980A) is used to measure the parasitic capacitance and inductance. The four-terminal ac resistance standard set (Agilent Model: 16074A) is used for the proposed work. The image and block diagram of the measurement setup are shown in Fig. 2.

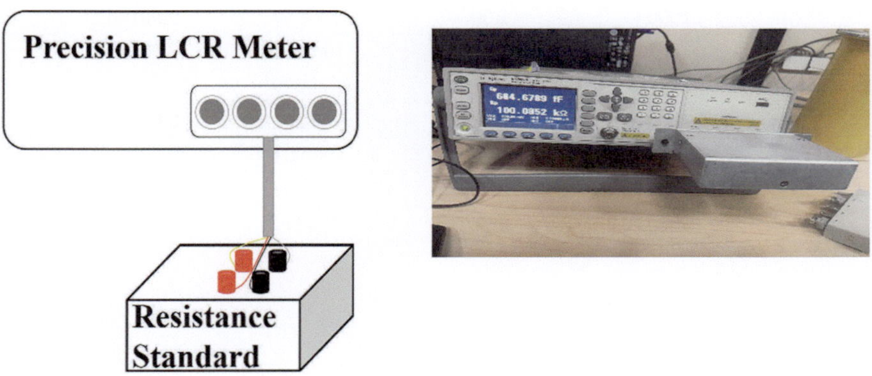

Fig. 2 Measurement setup for the measurement of parasitic components such as inductance and capacitance as per EECM

Measurement Procedure:

The following procedures are included in the measuring process to determine the metrological traceability of ac resistance:

1. Using a precision LCR meter to measure the parasitic component (capacitance or inductance) at 1 kHz as specified in the electrical equivalent circuit model.
2. Using Eqs. (2) and (3), calculate AC resistance for high and low value resistance, respectively
3. Calculate uncertainty in measurement for specified frequency from Eqs. (4) and (5).

The calculated resistance using Eqs. (2) and (3) and measured resistance using LCR Meter from 10 kHz to 1 MHz are compared as shown in Table 1. It is observed that the calculated and measured resistance are in good compliance.

Table 1 Comparison of calculated and measured AC resistance along with associated uncertainty in measurement

Nominal resistance value	Frequency, kHz	Calculated resistance, Ω	Measured resistance, Ω	± Exp. Unc., Ω
10 Ω	10	9.61230	9.61326	0.00123
	100	9.61230	9.61267	0.00145
	1000	9.61248	9.60149	0.03999
100 Ω	10	100.0507	100.0560	0.00529
	100	100.0507	100.0544	0.00624
	1000	100.0513	100.0110	0.69073
Nominal resistance value	**Frequency, kHz**	**Calculated resistance, kΩ**	**Measured resistance, kΩ**	**± Exp. Unc., kΩ**
1 kΩ	10	1.00048	1.00053	0.00004
	100	1.00048	1.00040	0.00004
	1000	1.00047	0.99968	0.00196
10 kΩ	10	10.0070	10.0071	0.0002
	100	10.0068	10.0039	0.0006
	1000	9.9893	9.9835	0.0503
100 kΩ	10	100.085	100.081	0.004
	100	99.911	100.003	0.003
	1000	85.748	99.121	0.003

6 Conclusion and Summary

The metrological traceability for AC resistance measurements is crucial for maintaining accuracy and reliability in electrical metrology. The frequency characterization of four-terminal pair resistance set ranging from 10 Ω to 100 kΩ has been done upto 1 MHz. This work will bridge the gap for AC resistance standards from audio to radio frequency. The proposed technique may be used to estimate the frequency characteristics of resistance standards upto MHz spectrum.

The proposed technique may be explored for the other type of AC/DC resistance standards in future.

Acknowledgements The authors are thankful to the Director, CSIR-National Physical Laboratory, New Delhi for his constant support and encouragement to carry out the proposed work.

References

1. Satish, Babita, Khurana B, Kumar S, Saxena AK (2015) Evaluation of four-terminal-pair capacitance standards using electrical equivalent circuit model. Measurement 73:121–126. https://doi.org/10.1016/j.measurement.2015.05.003
2. Satish, Sawhney N, Kumar S, Saxena AK (2015) Evaluation of three terminal capacitance standards at CSIR-NPL. Mapan-JMSI 30(4):261–265. https://doi.org/10.1007/s12647-015-0143-3
3. Agilent Technologies (2009) Agilent impedance measurement handbook, 4th ed

Realization of Sub-pF Capacitance Standards at Audio Frequency

Somya Dixit, Anshul, Aditi Sikarwar, Sachin Kumar, Jyotsana Mandal, Priyanka Jain, and Satish

Abstract The realization of a sub-pF capacitance standard is challenging, especially at audio and radio frequencies, which are critical for the semiconductor industry, serving as the backbone for electrical and electronic systems. In the proposed work, we demonstrate the realization of a sub-pF capacitance range using a dielectric test fixture based on parallel plates with guarded electrodes. Capacitance ranging from 100 fF to 1 pF is addressed for 1 kHz. Subsequently, capacitance is measured using a precise LCR meter, and the capacitance bridge, measured capacitance aligns well with theoretical capacitance. The uncertainty in measurement for the capacitance standards has been computed according to the guidelines outlined in the GUM document.

1 Introduction

The realization of sub-pF capacitance standards is crucial for various applications in the semiconductor industry, impedance measurements, and instrument calibration. Capacitive sensors are used in a variety of applications, and in some cases, changes in the sensor's capacitance due to external or internal stimuli are found to be in the range of femto-farads. This necessitates the development of accurate, precise, and reliable sub-pF capacitance standards to ensure the precise measurement of sub-pF capacitance variations.

The realization of sub-pF capacitance using the parallel plate method is a challenging task due to the inherent limitations of the parallel plate capacitor design. While a parallel plate capacitance with guarded electrode can be used to realize the sub-pF capacitance. Guarded electrode mechanism reduces the uncertainty in the capacitance measurement due to fringing field [1]. Generally, the dielectric test

S. Dixit · Anshul · A. Sikarwar · S. Kumar · J. Mandal · P. Jain · Satish (✉)
Electrical and Electronics Metrology Division, CSIR-National Physical Laboratory, New Delhi, India
e-mail: satish.nplindia@csir.res.in

S. Dixit · Anshul · A. Sikarwar
Banasthali Vidyapith, Newai, Rajasthan, India

S. Yadav et al. (eds.), *Advances in Metrology*, Lecture Notes in Mechanical Engineering,
https://doi.org/10.1007/978-981-96-6418-4_20

fixture with guard electrode is being used to measure the dielectric constant of solid, soft, flexible and fragile dielectric materials using contact and non-contact methods [2, 3]. Since the dielectric constant of air is known with very good precision and its properties are not deviating significantly as a function of frequency. In the proposed work, a dielectric test fixture with guarded electrode is utilized for the realization of sub-pF capacitance range from audio to radio frequency.

1.1 Theoretical Background

Insulating materials, also known as dielectrics, can be categorized as polar or non-polar based on their response to an applied electric field. When placed in an external field, they become polarized, meaning their internal charge distribution changes according to the applied field. Dielectrics have significantly fewer free-charge carriers compared to semiconductors and conductors. The complex dielectric constant, which depends primarily on frequency and temperature, provides information on a dielectric material's ability to store and dissipate electrical energy.

A parallel plate capacitor is usually separated by an insulating dielectric layer sandwiched between upper and lower electrodes [4]. In theory, the capacitance of a parallel plate capacitor is given by the Eq. (1):

$$C = \frac{\varepsilon A}{d} \tag{1}$$

where,

$$\varepsilon = \varepsilon_0 \varepsilon_r$$

where, ε_r represents the dielectric constant of the material, ε_0 vacuum permittivity is, A is the effective area of upper and lower plates, and d is the thickness or spacing between two electrodes. It can be seen that the capacitance is affected by three factors: the surface area of the two conductive plates (A), the distance or gap d between the two electrodes (d), and the relative permittivity of the middle dielectric layer (ε).

2 Measurement Setup and Observations

A dielectric test fixture (Agilent, Model: 16451B, Electrode Type B) and a precision LCR meter (Agilent Model: E4980A) is used to realize the sub-pF capacitances (Fig. 1 and Table 1).

In Fig. 2, the inverse of measured capacitance from lcr meter and capacitance bridge is plotted as a function of electrode separation. The measured capacitance has been fit using the linear function and it shows good correlation.

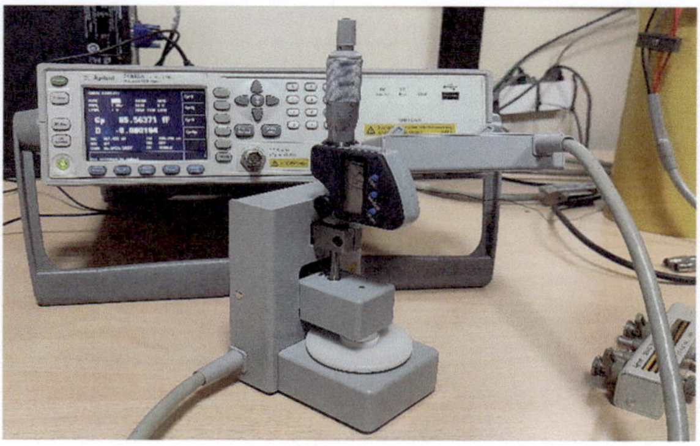

Fig. 1 Measurement setup for low value capacitance measurement facility

Table 1 Measured (Clcr and Ccb) and theoretical capacitance (Ct) with respect to electrode separation

Electrode separation, mm	Theoretical capacitance Ct, pF	± UCt, pF	Measured capacitance Clcr, pF	± UClcr, pF	Measured capacitance Ccb, pF	± UCcb, pF
0.2	0.86977	0.00440	0.8455	0.0094	0.8528	0.01171
0.5	0.34791	0.00075	0.3568	0.0023	0.3599	0.00192
0.6	0.28992	0.00054	0.2983	0.0017	0.3017	0.00129
0.7	0.24851	0.00041	0.2576	0.0013	0.2466	0.04475
0.8	0.21744	0.00032	0.2264	0.0011	0.2288	0.00086
0.9	0.19328	0.00026	0.2020	0.0012	0.2041	0.00075
1	0.17395	0.00022	0.1821	0.0011	0.1841	0.00058
1.5	0.11597	0.00012	0.1221	0.0009	0.1238	0.00027
2	0.08698	0.00008	0.0923	0.0010	0.0934	0.00017
2.5	0.06958	0.00006	0.0736	0.0012	0.0750	0.00013
3	0.05798	0.00005	0.0609	0.0018	0.0627	0.00010
3.5	0.04970	0.00004	0.0525	0.0014	0.0540	0.00009

3 Uncertainty in Capacitance Measurement

The capacitance of a parallel plate capacitor is inversely proportional to the distance between the plates and directly proportional to the area of the plates and the permittivity of the dielectric material between them. The uncertainty in capacitance measurement may be defined as:

$$\Delta c = \sqrt{(c_1 \Delta r)^2 + (c_2 \Delta d)^2} \tag{2}$$

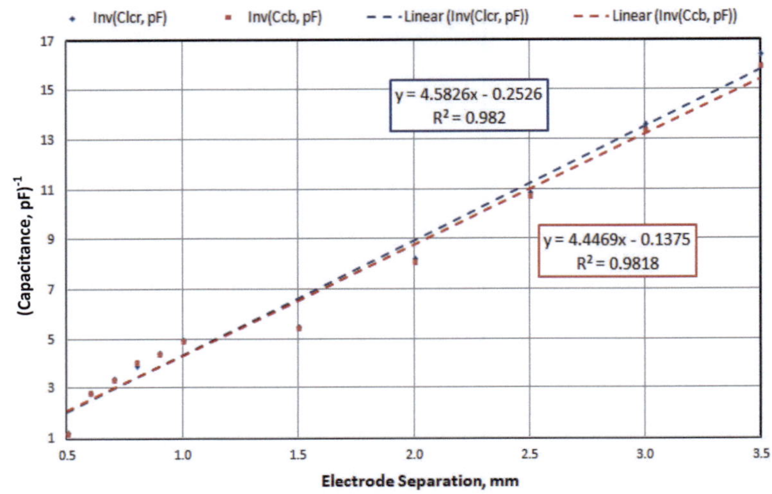

Fig. 2 Linearity of inverse of measurement capacitance as a function of electrode separation

$$C_1 = \frac{\partial C}{\partial r} = \frac{2\varepsilon_0\varepsilon_r\pi r}{d} \tag{3}$$

$$C_2 = \frac{\partial C}{\partial d} = -\varepsilon_0\varepsilon_r\frac{\pi r^2}{d^2} \tag{4}$$

$\Delta d = 0.002\,mm$, Uncertainty in electrode separation
$\Delta r = 0.005\,mm$, Uncertainty in radius
Where C_1 and C_2 are the sensitivity coefficients respectively

4 Summary

This work demonstrates the realization of sub-pF capacitance standards using a dielectric test fixture with guarded electrodes. The achieved capacitance range from 100 fF to 1 pF at 1 kHz is in compliance with theoretical capacitance values, highlighting the effectiveness of the proposed approach. The measured capacitance using precise LCR meter and capacitance bridge are found to be consistent with theoretical capacitance. The uncertainty in measurement has been computed using uncertainty propagation as per the GUM document. This work may be extended for the realization of sub-pF capacitances upto 1 MHz in future.

References

1. Keysight 16451B dielectric test fixture (2008). Keysight Technologies
2. Agilent Technologies (2006) Basics of measuring the dielectric properties of materials. Application Note. 5989–2589EN
3. Standard test method for dissipation factor and permittivity (Dielectric Constant) of Mica 1 (2017). ASTM. https://doi.org/10.1520/D1082-17
4. Grove TT, Masters MF, Miers RE (2005) Determining dielectric constants using a parallel plate capacitor. Am J Phys 73(1):52–56. https://doi.org/10.1119/1.1794757

Temperature-Dependence of Precise Quadrifilar Resistance Standard at Audio Frequency

Aditi Sikarwar, Somya Dixit, Anshul, Akash Kumar, Sachin Kumar, Jyotsana Mandal, Priyanka Jain, R. S. Meena, Suraj Khanna, and Satish

Abstract The proposed work describes the development of a calculable resistor also known as quadrifilar resistance, designed for use as an AC resistance standard, particularly at audio and radio frequencies. This resistor exhibits frequency dependence due to parasitic inductance and capacitance resulting from its construction and geometry. A quadrifilar resistance has been designed, fabricated and characterized for its temperature dependence and stability at 1 kHz using LCR Meter.

1 Introduction

A quadrifilar resistor is a special type of calculable resistor used as a precise reference standard for AC resistance [1]. Quadrifilar design is ideal for high-value calculable AC resistances, requiring high precision and stability at audio frequencies. The construction of a quadrifilar resistor involves folding the resistive element in specific ways to achieve a more compact design compared to traditional coaxial resistors. The resistive element is formed into a double loop, with both terminals positioned on the same side of the resistor, reducing its overall length [2, 3]. One key feature is that the current flows in opposite directions in its two halves, minimizing magnetic fluxes and improving stability and precision. A metallic shield will further reduce external electromagnetic interference, enhancing precision. Overall, the quadrifilar resistor offers a high precision and stability, making it a reference resistance standard for precise electrical measurements at low and audio frequencies, especially for higher resistance values where traditional coaxial resistors would be impractical. The

A. Sikarwar · S. Dixit · Anshul · A. Kumar · S. Kumar · J. Mandal · P. Jain · R. S. Meena · S. Khanna · Satish (✉)
Electrical and Electronics Metrology Division, CSIR-National Physical Laboratory, New Delhi, India
e-mail: satish.nplindia@csir.res.in

A. Kumar
Academy of Scientific and Innovative Research, Ghaziabad, UP, India

A. Sikarwar · S. Dixit · Anshul
Banasthali Vidyapith, Newai, Rajasthan, India

S. Yadav et al. (eds.), *Advances in Metrology*, Lecture Notes in Mechanical Engineering,
https://doi.org/10.1007/978-981-96-6418-4_21

temperature dependence of impedance parameter is one of the essential components of uncertainty in measurement [4, 5].

2 Measurement Setup and Observations

Calculable resistors are a type of resistor where the resistance value can be determined by simply measuring the physical dimensions of the resistor. A quadrifilar resistance standard has been designed and developed as a reference AC resistance standard as shown in Figs. 1 and 2. A four-terminal adapter cable (Agilent 16048A Test Lead) is used to connect the LCR meter to quadrifilar resistance. The image of the measurement setup to measure the quadrifilar resistance is shown in Fig. 3.

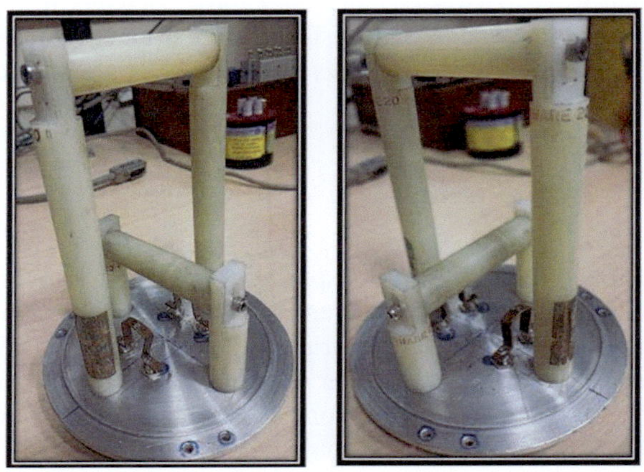

Fig. 1 Developed quadrifilar type resistance

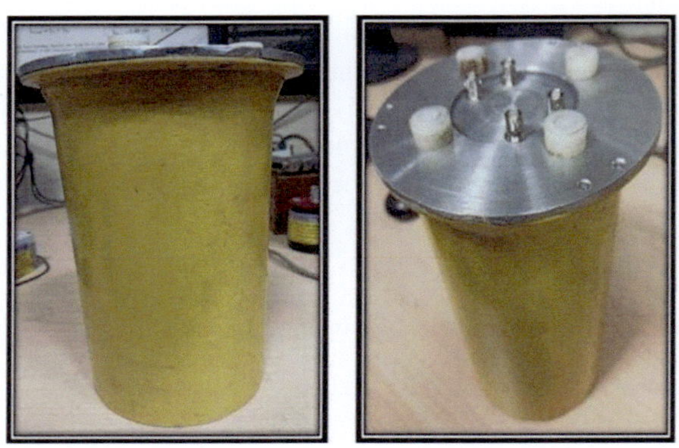

Fig. 2 Shielded quadrifilar resistance

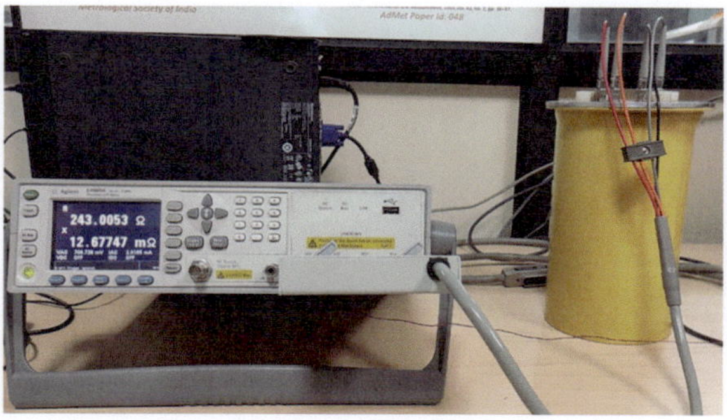

Fig. 3 Measurement setup for the quadrifilar resistance

The quadrifilar resistor is connected to an LCR meter (Agilent Model: E4980A) to measure the AC resistance using the IEEE 488.2 interface (GPIB), while a Thermo-hygrometer (Fluke Model:1620A) records the ambient temperature and relative humidity via the RS232 interface simultaneously as shown in Fig. 4. A measurement automation program has been developed on the LabVIEW platform to characterize the temperature-dependence of the quadrifilar resistor [6–8]. This automation program record ac resistance, temperature and relative humidity simultaneously. The front panel of the automation program is shown in Fig. 5.

The temperature dependence of the quadrifilar resistance is measured at 1 kHz using a precise LCR meter. The measured resistance exhibits a high positive-temperature-coefficient (PTC) as shown in Fig. 6. The measured resistance closely follows changes in temperature, with a response time observed less than 20 min.

The temperature coefficient for the quadrifilar resistance is found to be approximately 100–150 ppm as shown in Fig. 7, which is not desirable for reference resistance standards since they require minimal temperature-dependence.

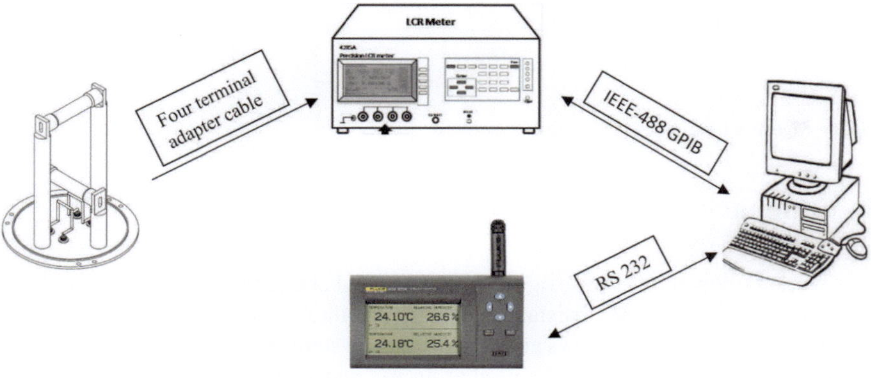

Fig. 4 Block diagram of measurement setup

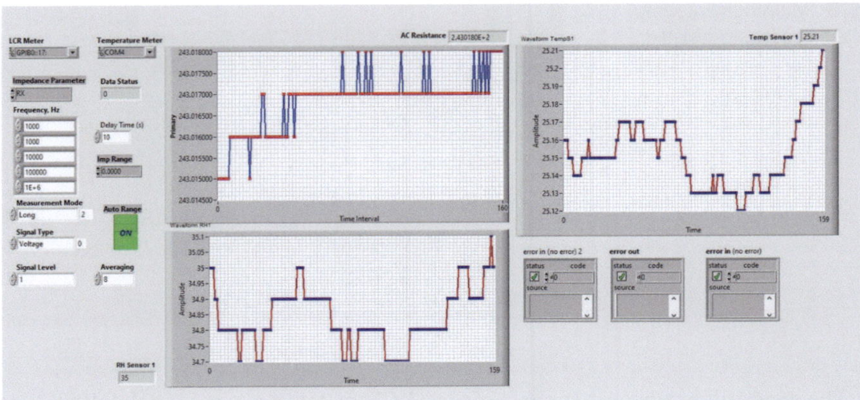

Fig. 5 Front Panel of measurement automation program

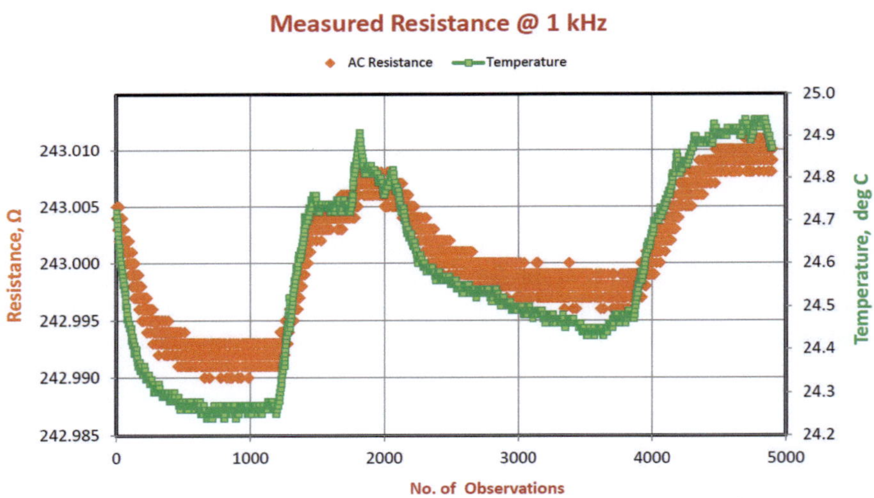

Fig. 6 Measured AC resistance with temperature

3 Summary and Discussion

This work demonstrates the design, fabrication, and characterization of a quadrifilar resistance for use as an AC resistance standard at audio and radio frequencies. While preliminary investigation shows good precision and stability at 1 kHz. The quadrifilar resistance exhibits promising potential as an AC resistance standard.

Our quadrifilar resistance standard exhibits excellent precision and stability at 1 kHz, making it a promising candidate for a reliable AC reference resistance standard. However, for practical application, future refinements could utilize an alloy

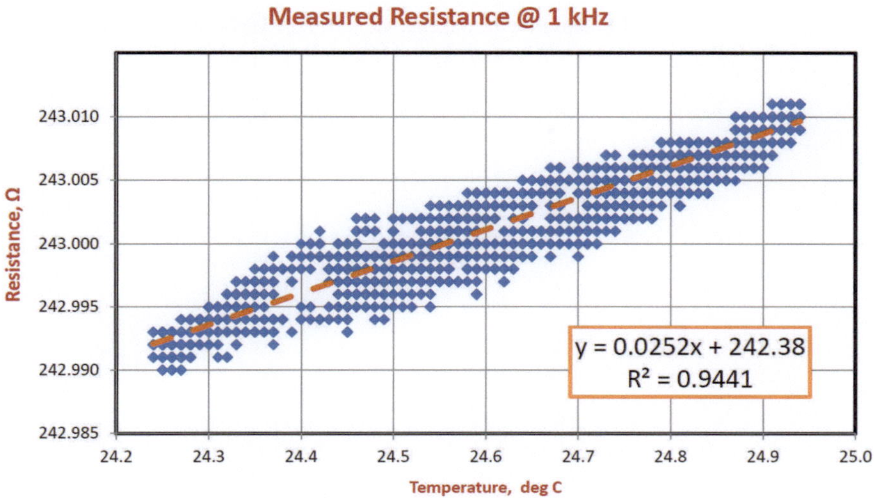

Fig. 7 AC resistance as a function of temperature at 1 kHz

with a lower temperature coefficient of resistance (TCR) to minimize temperature dependence and achieve a decade resistance value (e.g., 1 kΩ, 10 kΩ). AC reference standards typically employ decade resistance values to ensure compatibility and ease of comparison with existing primary and secondary standards.

Acknowledgements The authors are thankful to the Director, CSIR-National Physical Laboratory, New Delhi for his constant support and encouragement to carry out proposed work.

References

1. Agilent Technologies (2009) Agilent impedance measurement handbook, 4th ed
2. Pacheco-Estrada AH et al (2020) A simple methodology to develop bifilar, quadrifilar, and octofilar calculable resistors. https://doi.org/10.3390/app10051595
3. Bohacek J, Horska J, Sedlacek R (2010) Comparison of frequency dependences of resistance standards made from surface mount resistors. In: CPEM digest (conference on precision electromagnetic measurement), pp 647–648. https://doi.org/10.1109/CPEM.2010.5544765
4. Satish, Ansari MA, Saxena AK (2012) Determination and comparison of temperature coefficients of standard inductors using different methods. In: Conference on precision electromagnetic Measurements. IEEE, pp 404–405. https://doi.org/10.1109/CPEM.2012.6250974
5. Moreno JA (2016) Determination and use of temperature coefficients of inductance standards. In: CPEM 2016—conference on precision electromagnetic measurements, conference digest
6. Singh S, John T (2017) Automated module for characterization of reference standards of capacitance using impedance-matrix method. Turkish J Electr Eng Comput Sci 25:4142–4148. https://doi.org/10.3906/elk-1611-39
7. Satish, Babita, Khurana B, John T (2017) Measurement automation to implement evaluation procedure of four-terminal-pair capacitance standards using s-parameters. Mapan-JMSI 32(3):175–181. https://doi.org/10.1007/s12647-017-0211-y

8. Babita, Sharma DK, Satish, Ansari MA, Saxena AK (2014) A versatile automation program using LabVIEW for low dc current measurement. J Sci Ind Res (India) 73(2):91–94. Available: ISSN 224456

Improvement in Measurement Uncertainty for DC Voltage (Source) from 10 mV to 1000 V and Its Automation

Shreeyansh Golhani and Rahul Agarwal

Abstract This paper describes the measurement system established for DC Voltage ranging from 10 mV to 1000 V at SAC ISRO Ahmedabad. This was done by characterising the Multifunction calibrator using three standards i.e. 10 V DC reference standard cell, Precision DC voltage divider and Reference multimeter for DC Voltage Measurement. For this, a 10 V DC reference standard cell was used as the reference voltage source, a Voltage divider was used in comparison mode and a reference multimeter was used to measure the output voltage produced by a voltage divider. The 10 V DC reference standard cell used in this measurement was traceable to the primary standard Josephson junction at National Physical Laboratory New Delhi. This measurement process was also done through automated mode. Expanded Measurement Uncertainty was calculated at coverage factor $k = 2$ and achieved measurement uncertainty is 3.14 ppm @ 10 mV and 2.34 ppm @ 1000 V.

Keywords Null detector · Voltage divider · Reference multimeter

1 Introduction

Calibration laboratory at SAC ISRO Ahmedabad is a central laboratory, which provides calibration services in the frequency range from DC to 50 GHz. In the field of DC parameters, multifunction calibrators Fluke 5730A/5720A are used as standard laboratory equipment which provide DC output from 10 mV to 1000 V [1]. For calibration of these multifunction calibrators, three standards are used: 10 V DC reference standard cell (Fluke 732B) which is calibrated from higher echelon laboratory, Precision DC voltage divider (Guildline 7520A) and Reference multimeter (Fluke 8508A). The Fluke 732B is a DC Zener voltage reference which produces two outputs, 1.018 and 10 V. The 10 V DC output of this reference standard cell was used as the reference voltage source [2]. The Guildline 7520A Precision DC voltage divider is the first fully automated Self-Calibrating sub-ppm voltage divider [3]. It

S. Golhani (✉) · R. Agarwal
Space Applications Centre, ISRO Jodhpur Tekra Ahmedabad, Gujarat, India
e-mail: shreeyansh@sac.isro.gov.in

S. Yadav et al. (eds.), *Advances in Metrology*, Lecture Notes in Mechanical Engineering,
https://doi.org/10.1007/978-981-96-6418-4_22

provides ratios of 1:1, 10:1, 100:1 and 1000:1 with input voltages upto 1100 V_{dc}. The 7520 has two modes of operation, Comparator Mode (i.e. compares two separate sources) and Divider Mode (i.e. divides the output from a voltage reference such as a zener or a calibrator) [4]. The 7520 was used in comparator mode and the reference multimeter was used to measure the output voltage produced by the voltage divider. When the reference multimeter is used as a Null detector [5] to measure the 7520 output voltage, there are three errors that can occur with the reference multimeter: the input bias current of the meter, loading effect of the DMM due to the input impedance of the meter, and a relatively noisy measurement when connected to a device that has more than 100 kΩ output impedance. To obtain valid results, all these errors must be considered in the setup and use of the 7520 voltage divider. Only then can a reference multimeter be used to get good results. With proper configuration of the measurement setup and correct procedures, these errors can be greatly reduced.

2 Methodology

Calibration of the multifunction calibrator is performed using a DC reference cell, precision voltage divider and reference multimeter. The precision voltage divider is used in comparator mode and the difference between the calibrator output voltage and the reference voltage is measured with the multimeter used as a null detector [6]. In comparator mode, various range selections are present from 10 mV range to 1000 V range. To calibrate the decade voltage output of the multifunction calibrator, the same range has to be selected on the precision voltage divider. Each measurement of decadal voltage output was performed in two steps. In the first step, measure the DC bias error of the Null detector because for low uncertainty measurements [6], accounting for the bias current is necessary to ensure accurate readings, especially for high impedance inputs. If the voltage divider has an output impedance that is in the tens of kΩ or higher, the DMM bias current could cause offset errors of several microvolts which is larger than the accuracy specification of the 7520 Voltage Divider. For this first short the reference voltage and voltage input terminals of the 7520 divider as shown in Fig. 1. Now measure the Null voltage produced by 7520 precision voltage divider using the Null detector/Multimeter. This voltage is considered as the DC Bias error of the Null detector (i.e. multimeter). And this voltage is measured at the different voltage divider ranges at +10 V reference as shown in the below chart Fig. 2.

In the second step, the Reference cell's 10 V output and multifunction calibrator's output are Connected to the reference input and voltage input of voltage divider respectively through low thermal EMF cable as shown in Figs. 3 and 4. Select Precision voltage divider in the comparator mode with the same range as the output present in the multifunction calibrator and use the +10 V Ref level. Now apply the required DC voltage from the multifunction calibrator. For voltages >10 V the precision voltage divider in the comparator mode divides the calibrator voltage down to 10 V and the difference between the divided calibrator voltage and the 10 V reference is

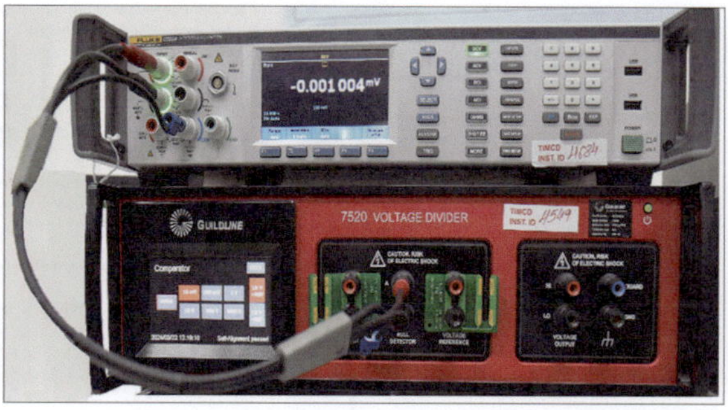

Fig. 1 Setup for DC Bias voltage measurement

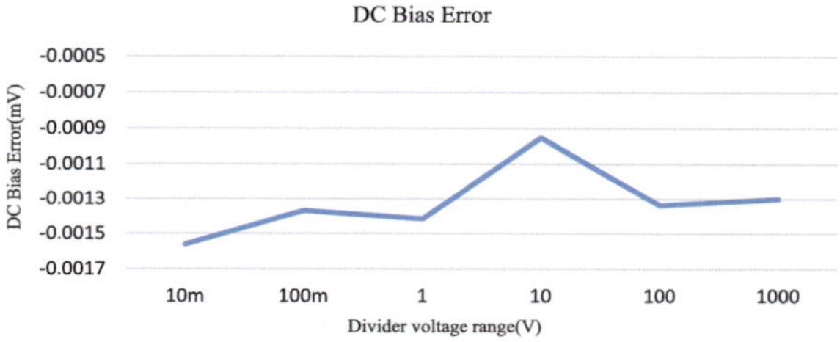

Fig. 2 DC Bias voltage error w.r.t different input voltage range

measured with the Null detector (i.e. reference multimeter). For voltages <10 V the difference between the divided 10 V reference and the calibrator output voltage is measured by the Null detector. For negative voltage measurement set divider in −10 V Ref level and apply the required negative voltage from the calibrator. All observations from the Null Detector are taken for measurement uncertainty calculations.

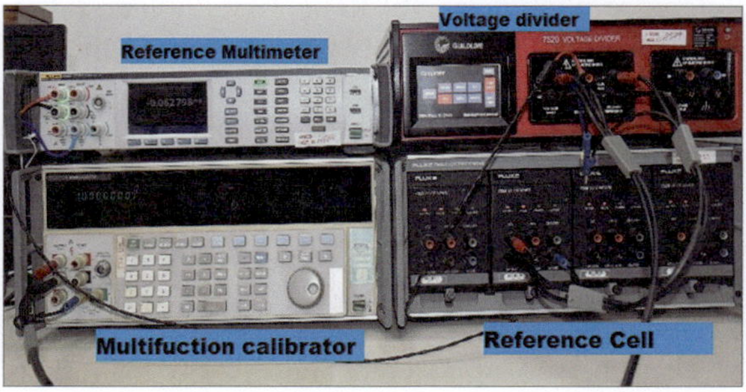

Fig. 3 Setup for DC voltage measurement with divider in comparator mode +10 V reference

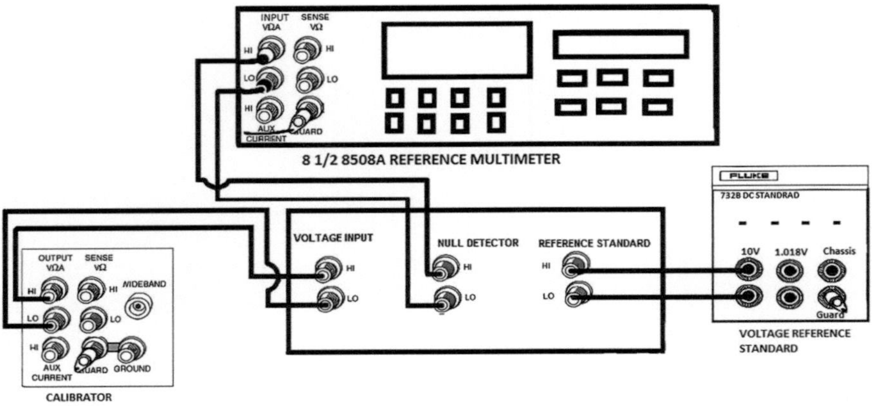

Fig. 4 Block diagram for DC voltage measurement with divider in comparator mode

3 Discussion

Measurement Uncertainty was calculated using the observations for all DC voltage points in the range from 10 mV to 1000 V, considering coverage factor k = 2. Measurement uncertainty consists of two parts Type A and Type B uncertainty [7]. Type-A uncertainty covers the random uncertainty part of measurement and Type B uncertainty covers the systematic error part during the measurement.

Type A uncertainty was calculated based on the observations by taking the standard deviation of all observations.

Now Type B uncertainty was calculated considering different systematic errors present during the measurement. Type-B affecting factors are: (a) Resolution of Reference multimeter, (b) floor value in each range of Reference multimeter, (c) stability of divider, (d) dc bias voltage, (e) calibration uncertainty of dc reference cell.

All these systematic errors were divided by the corresponding probability distribution function.

Now the combined uncertainty was calculated by taking the Root Sum Square of type-A uncertainty and all type-B uncertainty components. This combined uncertainty is multiplied by the coverage factor to get expanded uncertainty. Measurement Uncertainty budget is tabulated in Table 1.

The achieved measurement uncertainty for various ranges is shown in Fig. 5 and the values are tabulated in Table 2.

Table 1 Measurement uncertainty budget

Sr. No	Quantity/affecting parameter	Type A / Type B	Sensitivity coefficient	Probability distribution function	Degree of freedom	Uncertainty of each component
Random uncertainty component						
(a)	Taking observations and calculate Standard Deviation	Type A	1	Normal	No. of observations	Type A
Systematic uncertainty component						
(a)	Resolution of Reference multimeter	Type B	1	Rectangular	∞	Type B1
(b)	Floor value of Reference multimeter	Type B	1	Rectangular	∞	Type B2
(c)	Stability of divider	Type B	1	Rectangular	∞	Type B3
(d)	DC bias voltage	Type B	1	Rectangular	∞	Type B4
(e)	Calibration uncertainty of dc reference cell	Type B	1	Rectangular	∞	Type B5
Combined uncertainty = SQRT {$(\text{Type A})^2$ + $(\text{Type B1})^2$ + $(\text{Type B2})^2$……. + $(\text{Type B5})^2$}						
Expanded uncertainty = Combined Un * Coverage factor (K = 2 for 95.46% confidence level)						

Table 2 Measurement uncertainty for all range

DC Output voltage range (V)	Measurement uncertainty (ppm)
10 m	3.14
100 m	2.35
1	2.33
10	2.83
100	2.33
1000	2.34

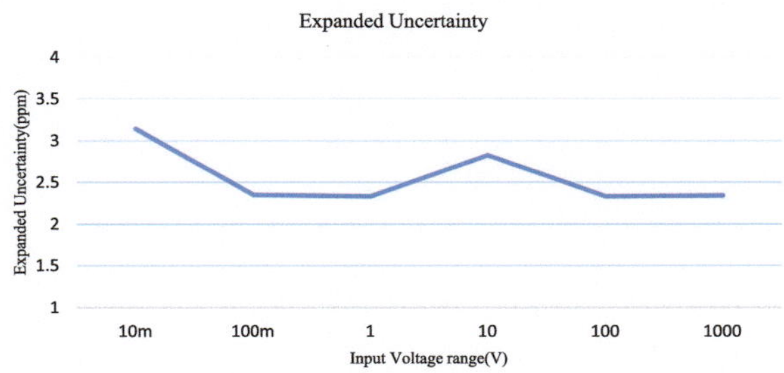

Fig. 5 Expanded uncertainty for DC voltage source from 10 mV to 1000 V

4 Conclusion

The two steps calibration process described in this paper is suitable for calibration of multifunction calibrators which provide DC voltage ranges from 10 mV to 1000 V. All measurements are traceable to the 10 V standard cell. This measurement process was also done through automated mode and completed within 3–4 hours. Expanded Measurement Uncertainty was calculated at coverage factor $k = 2$ and achieved measurement uncertainty is 3.14 ppm @ 10 mV and 2.34 ppm @ 1000 V. For full range from 10 mV to 1000 V.

References

1. Fluke 5700A/5720A Series II calibrator Service manual, Fluke Corp
2. 732A DC Reference standard Technical data, Fluke Corp
3. Operation and instruction manual, Guildline 7520
4. Best practices using the guildline 7520 automated voltage divider, Guildline Instruments
5. Replacing analog null detectors with precision multimeters, application note. Fluke Corp
6. Using Digital Multimeters in place of analog null detector for metrological applications, Fluke Corp
7. Guidelines for estimation and expression of uncertainty in measurement, Application note NABL-141

Current Challenges in the Techniques Used for Particulate Matter Monitoring Under the Indian Conditions

Ashmeet Kaur Alang and Shankar G. Aggarwal

Abstract With current science and policy needs, more attention needs to be given to better measurement protocols of air pollutants which is crucial for evaluating the overall air quality. Air monitoring stations managed by institutions or government agencies for monitoring air pollutants are usually point-based and location-specific, with most of them located in urban regions. Given the heterogeneous sources of aerosols in different seasons and locations in the Indian region, the spatial coverage of such measurements is insufficient. In the recent past, many studies have been conducted which utilized satellite-derived reanalysis PM concentration for assessing the air quality over a wide spatial distribution. However, what the satellite-derived reanalysis data make up by obtaining continuous spatial distribution of PM, they lack in accuracy of the measured data due to several limitations discussed in detail in the present work. The bias in the accurate reanalysis PM concentration dataset is especially evident when there is very high loading of PM in post-monsoon and winter season in the Indo-Gangetic Plain. Compared to the commonly used methods for PM monitoring, utilization of columnar concentration measurement techniques such as Multi-Axis Differential Optical Absorption Spectroscopy (MAX-DOAS) appears to be a suitable method. Despite being an important contributor to air quality, there is a lack of standardization of PM mass measurements in India. It has been noted that standardizing the measurement techniques and parameters is urgently needed.

Keywords Differential optical absorption spectroscopy (DOAS) · Representative PM distribution · Standardized PM measurement techniques · Beta attenuation method · Remote sensing PM monitoring

---

A. K. Alang · S. G. Aggarwal (✉)
Academy of Scientific and Innovative Research (AcSIR), Ghaziabad, India
e-mail: aggarwalsg.nplindia@csir.res.in

Environmental Sciences and Biomedical Metrology Division, CSIR-National Physical Laboratory (NPL), New-Delhi, India

S. Yadav et al. (eds.), *Advances in Metrology*, Lecture Notes in Mechanical Engineering,
https://doi.org/10.1007/978-981-96-6418-4_23

1 Introduction

Particulate matter (PM) is a crucial factor in assessing air quality, due to its detrimental impacts on human health [3, 17], visibility [38, 39], earth's ecosystem and climate change [16, 20]. PM concentration can be measured as PM_{10}, $PM_{2.5}$, or PM_1, based on the mass concentrations of particles with aerodynamic sizes below 10, 2.5, and 1 μm, respectively. Particularly, $PM_{2.5}$ can penetrate deep into the respiratory system causing cardiovascular and respiratory diseases. The size of PM is innately related to the inhalation and deposition of particles in the human respiratory system [27]. Considering this information, regulatory agencies responsible for safeguarding public health must comprehensively examine pollutants' trends, monitoring networks, and their measurement techniques for accurate evaluation of air quality, which in turn is important for formulating better control policies. Ideally, these monitoring networks will enable regulatory agencies to provide advisories and impose suitable restrictions in specific areas when pollutants concentrations rise to unhealthy levels.

For measuring PM concentration, most countries' Ambient Air Quality Standards (AAQS) specify methods such as manual sampler, tapered element oscillating microbalances (TEOM) and beta attenuation monitors (BAM) to ensure accurate and precise measurement of PM. Several studies which performed inter-comparison of these 3 techniques, have shown more than 30% variations in PM mass measurement [1]. It is also noteworthy that the temporal and spatial mass concentrations of PM can differ greatly within a region. The PM concentrations recorded by a single monitoring site might not accurately reflect the air quality of that region. Recently, this concern has become particularly pressing in developing countries undergoing industrialization [14, 33]. The Indo-Gangetic Plain (IGP), experiences high aerosol concentrations year-round. This region exhibits significant spatio-temporal variability in aerosol characteristics, including their seasonal and inter-seasonal fluctuations. The diversity in aerosol emission sources and compositions contributes significantly to this variability.

So far, the work done for standardised monitoring and measurement of PM in national metrological institutes (NMIs) and regulatory institutions worldwide is not up to satisfactory level. In the present work in the context of PM measurement, we discuss the commonly used methods for PM monitoring and mass measurements in India, identifying their technological gaps and limitations based on previously published results, and suggesting better PM measurement techniques.

2 Primary/Reference/Ground-Based Methods for PM Measurement

Majorly three methods have been mandated by the AAQS worldwide for the measurement of PM_{10} and $PM_{2.5}$ namely: (i) collection on filter using high- or low-volume air samplers over a 24 h span and subsequent gravimetric determination (offline primary method), (ii) collection on filter-tape and subsequent determination by the difference in the transmission of beta radiation (real-time online method), and (iii) collection on filter placed on top of oscillating element which oscillates at a frequency inversely proportional to the mass of collected PM (tapered element oscillation microbalance technique, online method) [1].

Although filter-based high- and low-volume air samplers have demonstrated reliability and accuracy in detailed studies of airborne particles during intensive experiments, they present several limitations when applied in routine monitoring networks for regulatory purposes. The primary drawback of filter-based samplers for PM measurement is their inability to provide real-time data and ineffective to observe peak concentration hours. It often takes weeks or even months from the time samples are collected until the PM data becomes available. This delay prevents regulatory agencies from responding promptly to the changes in the air quality.

The high- or low-volume air samplers are majorly classified based on flow rate. The Bureau of Indian Standards (BIS) IS 5182 (Part 23): 2006 specified flow rate of 1130 L per minute (lpm) or 67.8 m^3/h for high-volume air samplers, and 1 m^3/h for low-volume air samplers. When a low-volume sampler is operated for 24 h span, the maximum volume of air sampled would be ~24 m^3, which is a quite small or limited amount of air sampled. This can lead to less representative samples, especially in areas with high PM loading and large heterogeneity (Fig. 1). In addition, in regions with high PM levels, the smaller sampling capacity of low-volume air samplers may not be sufficient to capture the true extent of air pollution, leading to potential underestimation of PM concentrations. Furthermore, due to the lower flow rate, low-volume samplers require longer sampling periods to collect sufficient particulate matter for accurate analysis. This extended duration can delay the availability of data required for prompt action plans. These limitations make low-volume air samplers less ideal for accurately measuring PM concentrations in environments with large variability in pollution levels, such as many urban and industrial areas in India.

Therefore, even with the availability of the primary gravimetric method, there remains a need for a real-time automated technique that can (i) produce data directly correlated with mass, (ii) allow for the combination of data from various cities and seasons for analysis, (iii) minimize the handling of delicate filters, and (iv) enable short-term ambient particulate concentration measurements (1–8 h intervals) in the same units ($\mu g/m^3$) as long-term (24 h interval) samples [31]. The beta-ray attenuation method (BAM) is widely used for continuous real-time monitoring of ambient PM in several studies [8, 26, 25, 30]. Recently, Shukla and Aggarwal [30] checked the performance evaluation of BAM in different ambient conditions particularly

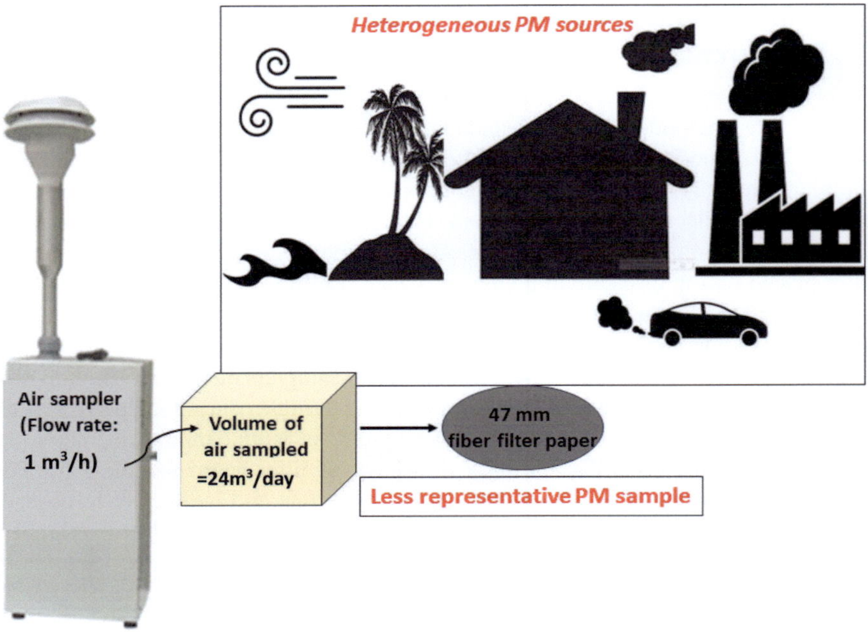

Fig. 1 Depiction of less representative PM samples obtained from a low volume air sampler

for high $PM_{2.5}$ episodes and relative humidity (RH) by comparing the data with collocated gravimetric sampling in Delhi, India. The findings indicate that BAM measurements align within 10% of the gravimetric method at low ambient $PM_{2.5}$ concentrations (below 130 μg/m^3) and low relative humidity (under 60%). However, as $PM_{2.5}$ concentrations and relative humidity increase, the discrepancy in BAM measurements can grow by up to five times. The study concludes that the accuracy of the beta gauge method is significantly influenced by relative humidity, and the data it produces is reliable only under conditions of low mass loading and low ambient humidity. Therefore, the beta attenuation method, while useful in certain conditions, is not ideally suited for monitoring PM in India, particularly due to the country's significant spatial heterogeneity in PM distribution.

India exhibits considerable spatial heterogeneity in PM distribution, with significant differences in particulate concentrations between urban and rural areas. However, the majority of ambient PM monitoring and measurement studies have been focused on larger cities, with fewer conducted in small cities and rural areas [2]. Moreover, significant differences in PM concentrations are observed even within closely located sampling locations of the same city. This may be due to different sampling locations and meteorological conditions, but to large extent variations are observed due to lack of standard operating methods [21]. To gain a more accurate understanding of PM concentration levels and trends, this imbalance needs to be addressed, as it is essential for comprehending the overall state of PM pollution in India.

3 Satellite-Derived Reanalysis Data for PM Monitoring

While ground-based air monitoring networks have traditionally provided localized, precise measurements, satellite-derived reanalysis data has emerged as a powerful complementary tool. Satellite data provides near-global coverage, which is especially beneficial in regions where ground-based monitoring stations are sparse or non-existent. This allows for the monitoring of PM levels in remote and rural areas, as well as across entire continents. It also allows for the comparison of PM levels across borders, which is critical for understanding transboundary pollution and coordinating international efforts to address air quality issues. Due to their continuous temporal and spatial resolution, high data quality and wide aerosol species, atmospheric reanalysis integrating satellite observations is a helpful tool to provide the most complete picture of global climate change and spatial environment [29].

Several satellite datasets are widely used for monitoring PM, offering valuable insights into global air quality. The Moderate Resolution Imaging Spectroradiometer (MODIS) aboard the Terra and Aqua satellites is a key source, providing Aerosol Optical Depth (AOD) data that serves as a proxy for surface PM concentrations [4, 22]. Similarly, the Multi-angle Imaging Spectro Radiometer (MISR) on Terra enhances aerosol characterization by delivering AOD data from multiple angles [9]. The Ozone Monitoring Instrument (OMI) on the Aura satellite, while primarily focused on ozone, also provides Aerosol Index (AI) and AOD data, detecting absorbing aerosols like black carbon [13]. The Visible Infrared Imaging Radiometer Suite (VIIRS) aboard Suomi NPP and NOAA-20 offers high-resolution AOD data, enabling day and night observations [5]. The Modern-Era Retrospective analysis for Research and Applications, Version 2 (MERRA-2) dataset, generated by NASA's Goddard Earth Observing System Model, assimilates satellite data to provide reanalysis products that include detailed information on aerosols and PM, making it essential for long-term climate and air quality studies.

While the reanalysis data provides valuable insights for monitoring PM, it also has several limitations. Satellite sensors primarily capture AOD, which indicates the total amount of aerosols present throughout the entire atmospheric column rather than surface-level concentrations. This makes it difficult to precisely estimate PM levels at ground level because the relationship between AOD and surface PM is complex and varies depending on factors like the vertical distribution of aerosols, weather conditions, and the specific types of aerosols present [9]. Further, the accuracy of satellite-derived data is often reduced due to cloud cover, which can obscure aerosols and create gaps in the data. Additionally, although satellite data is collected frequently, its temporal resolution may not be sufficient to capture rapid fluctuations in air quality, such as those caused by sudden pollution events [9].

Satellite-based observations offer broad spatial coverage but generally lack the fine resolution of ground-based monitoring networks. This limitation can lead to errors, particularly in densely populated urban regions where PM levels fluctuate significantly over small distances [36]. A recent study by Alang and Aggarwal [2] evaluated MERRA-2's reanalyzed $PM_{2.5}$ concentrations against ground-based measurements across various Indian cities from 2015 to 2019, considering seasonal and regional variations (Fig. 2). The findings indicate that MERRA-2 struggles to accurately represent $PM_{2.5}$ mass concentrations during peak pollution periods, as evidenced by an underestimation of 34% and 38% in the post-monsoon and winter seasons, respectively. This discrepancy is likely due to the absence of nitrate in MERRA-2's $PM_{2.5}$ dataset, a critical component of fine PM in the Indian atmosphere [10, 19]. Another factor contributing to these deviations in winter and post-monsoon seasons is MERRA-2's coarse resolution, which prevents it from capturing localized pollution sources in urban cores [28]. Similarly, [23] analyzed $PM_{2.5}$ concentrations in 20 Indian cities from 2015 to 2018 and found that while the observed average was 80 $\mu g/m^3$, MERRA-2 estimated only 35 $\mu g/m^3$, revealing an underestimation of 34%.

Satellite-derived reanalysis data is a powerful tool for providing broad, consistent, and temporally frequent monitoring of PM across large areas. However, it is best used in conjunction with ground-based monitoring networks, which offer higher resolution and direct measurements necessary for detailed, localized air quality assessments.

4 Light Detection and Ranging (Lidar) Sensors

Lidars are effective remote sensing tools that work in the optical wavelength spectral range with distinct abilities for atmospheric observations. Within this spectral range, the interactions between electromagnetic radiation and atmospheric components are significant, enabling the detection of aerosols and trace gas concentrations [7].

Lidar systems typically comprise co-located transmitter and receiver units, along with high-speed photodetectors and acquisition electronics. The basic setup of a lidar is shown in Fig. 3. For aerosol remote sensing, the transmitter is generally a pulsed laser and the receiver an optical assembly with a telescope which acts as a collector of the backscattered radiation. The optical signal is filtered to minimize noise from background radiation, then converted into an electrical signal using a fast photodetector. This signal is amplified, digitized, and stored for subsequent processing.

The benefit of lidar-based active remote sensing techniques lies in their capability to perform remote measurements to assess contaminant levels in the target area and quickly gather information on meteorological conditions, such as boundary layer height/thickness, wind speed, and atmospheric air temperature [24]. Elastic scattering lidar, Raman lidar, differential absorption lidar (DIAL), Doppler lidar, and fluorescence (resonance) lidar are among the most extensively used techniques for remote sensing in air monitoring studies. A previous study by Lewis et al. [18] compared the findings from two airborne aerosol lidar campaigns conducted over the San Joaquin

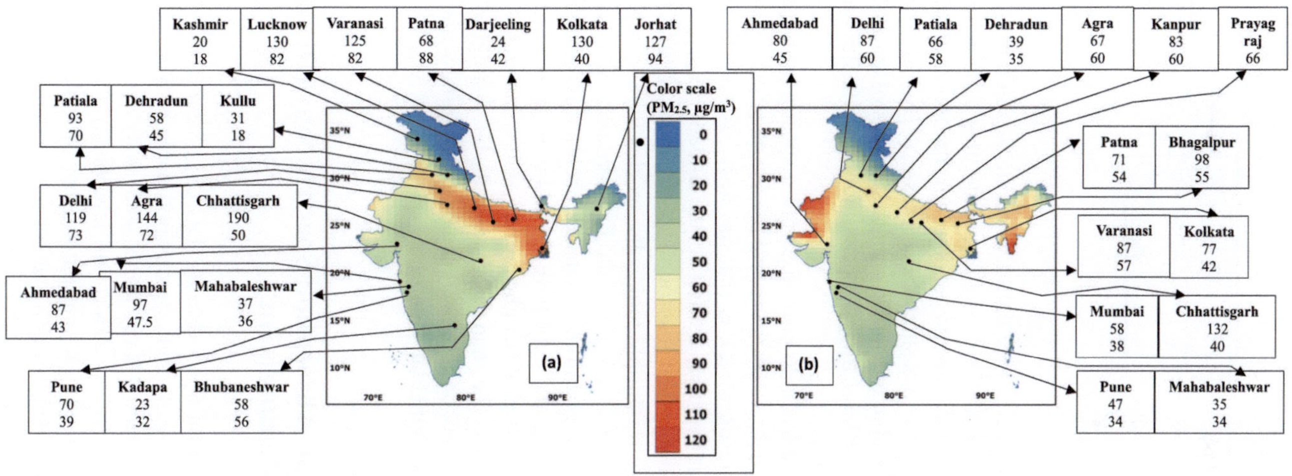

Fig. 2 Distribution of average $PM_{2.5}$ and location of measurement sites along with corresponding ground $PM_{2.5}$ and MERRA-2 $PM_{2.5}$ concentration during the time period 2015–2019, where **a** winter, **b** summer season. The upper values in boxes indicate ground $PM_{2.5}$ and the lower values indicate MERRA-2 $PM_{2.5}$ at the sampling location [2]

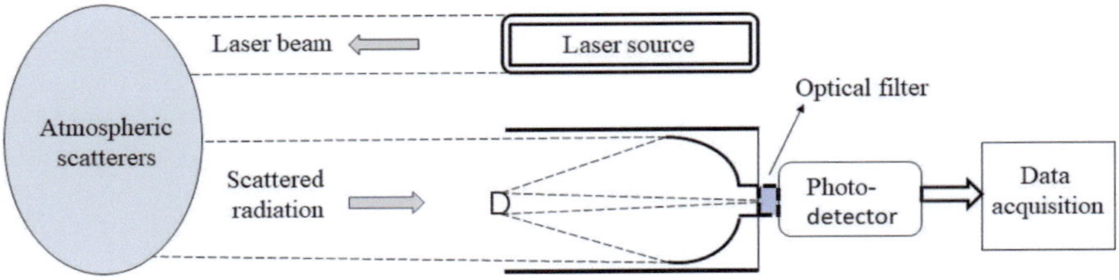

Fig. 3 Basic schematic of a lidar system

Valley, California and compared to MODIS (Moderate Resolution Imaging Spectroradiometer) AOD data from the Terra and Aqua satellites. Although satellite data provide a wide-area view of AOD, they do not offer information on the vertical distribution of aerosols, making lidar measurements necessary. By integrating lidar and satellite data, this study aimed to enhance the understanding of aerosol distribution and its connection to ground-level $PM_{2.5}$ concentrations for health and regulatory purposes across a wide region.

Techniques such as tunable diode laser absorption spectroscopy (TDLAS) and DIAL, utilise absorption spectroscopy to achieve precise quantification of molecular species. However, the main drawback is that, in most cases, each target species requires a different excitation laser, adding to the already high cost of components. In contrast, differential optical absorption spectroscopy (DOAS) utilizes UV/visible absorption to measure multiple molecular species with a single broadband light source, discussed in the following section. Furthermore, lidar systems require regular calibration and stable operation to ensure accurate measurements, which can be technically challenging and expensive. In addition, its performance can be affected by high humidity, dense clouds, or heavy rain, which can scatter or absorb the laser signal, reducing accuracy. They can detect aerosols, but distinguishing between aerosol types (e.g., dust, smoke, or pollution) can be challenging without additional spectral information or complementary sensors.

5 Differential Optical Absorption Spectroscopy (DOAS) Method for PM Measurement

Differential Optical Absorption Spectroscopy (DOAS) is an effective method that allows the monitoring of aerosols and gaseous pollutants utilising a combination of light emitters, light receivers, optical fibres, and analysers. The measurement principle is based on Beer-Lambert absorption law which states the relationship between the amount of absorbed light and the number of molecules that the light passes through. In DOAS, two absorption spectra are recorded: one as a reference scan through an atmospheric column ideally free of the analyte, and the other through the same pathlength containing the analyte. Then by subtracting these two spectra

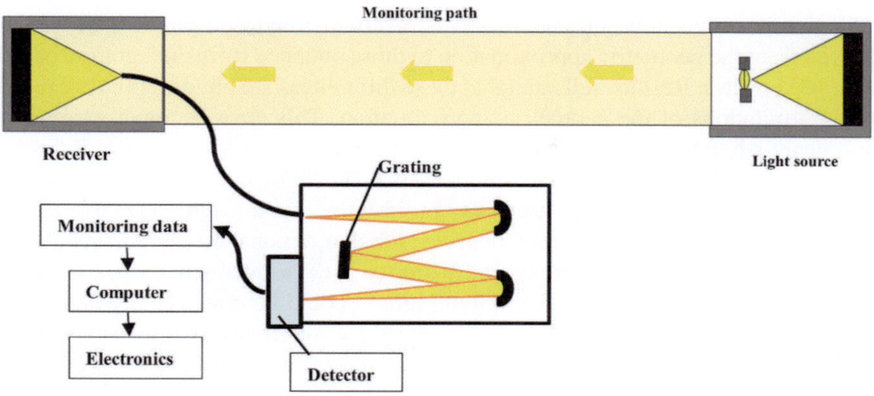

Fig. 4 DOAS horizontal column-measurement set-up for monitoring emissions

and applying the Beer-Lambert law, the resulting differential spectrum depends only on the absorption cross-section of the analyte species and the difference in column densities between the analyte and the reference spectrum [32]. Figure 4 shows a typical DOAS setup where, a concave mirror is utilized to collimate a broadband light source, usually a D2 lamp, across the monitoring path. A second concave mirror is then employed to focus the transmitted light into a fiber-optic cable, which subsequently channels it into the spectrometer. There are many advantages of using DOAS method. Wide range of species can be effectively measured with a single broadband light source. Also, it is a cost-effective method due to inexpensive Charge-Coupled Device (CCD) spectrometers and a constant current D2 emission lamp.

In addition to continuous emission monitoring systems, sunlight-based DOAS is also widely used for general environmental monitoring, i.e. multi-axis differential optical absorption spectroscopy (MAX-DOAS) technique used to measure trace gases and aerosols in the atmosphere. It is highly regarded for its capacity to deliver precise and real-time measurements across extensive spatial and temporal ranges by the determination of vertical distributions of trace gases and aerosols in the troposphere by observing scattered sunlight at various elevation angles [12, 37] using relatively simple and cheap ground-based instruments. Multiple MAX-DOAS networks have been established to conduct extended measurements. These long-term observations, along with data from short-term measurement campaigns, have been utilized in environmental research and for validating satellite data and model simulations at Tsukuba, Japan [15], Beijing, China [11] and Wuxi, China [35]. Thus, using different inversion approaches, the column densities, vertical profiles and near-surface concentrations of the trace gases and aerosols can be derived and provide additional information compared to in-situ monitoring or satellite observations.

Ground-based MAX-DOAS is a valuable technique for observing tropospheric aerosols, but it comes with certain limitations. The tropospheric vertical column density (VCD) can be determined either using the geometric approximation method [6] or by integrating the retrieved concentration profiles [34]. In a study by Wang et al.

[35], significant discrepancies are observed on comparing tropospheric VCDs from derived from the geometric approximation to those obtained by the integration of the retrieved profiles. Results indicated that these differences are almost entirely caused by the limitations of the geometric approximation. This error becomes especially significant when the aerosol load is strong, which is the situation in most industrialised regions. Further, the ground-based MAX-DOAS observations are limited to the specific locations where instruments are installed. As a result, data coverage is sparse and does not provide comprehensive global or regional measurements. Additionally, these measurements are typically limited to daytime observations due to the reliance on scattered sunlight. These limitations should be considered when interpreting MAX-DOAS data, and complementary measurement techniques are often used to address some of these challenges. Therefore, generally DOAS performed using sunlight is not advisable for continuous emission monitoring systems due to inherent fluctuations of sunlight levels.

6 Conclusion

Although significant progress has been made in the monitoring and measurement of particulate matter (PM) in India, the imperative for standardization in PM measurement remains critical and has not been adequately addressed. The significant spatial variability in PM composition and the high PM loading across different regions of the country pose unique challenges that current methodologies do not fully account for. Without standardized and regionally tailored measurement techniques, the accuracy of PM data remains compromised, hindering effective air quality management and public health protection. Addressing these gaps through rigorous standardization efforts is essential to improving PM monitoring and ensuring reliable data for policy-making. Based on the current discussions it is suggested that horizontal column-based measurement of aerosols is a better option for monitoring particulate matter under Indian conditions.

Acknowledgements The authors are thankful to Director, CSIR-NPL, New Delhi for all the support and encouragement. AK is thankful to University Grant Commission (UGC) for providing the fellowship under Project No. 90802 (UGC-SRF fellowship). Funding from OLP 210332 is also acknowledged.

References

1. Aggarwal SG, Kumar S, Mandal P, Sarangi B, Singh K, Pokhariyal J, Gupta PK (2013) Traceability issue in PM 2.5 and PM 10 measurements. Mapan 28:153–166
2. Alang AK, Aggarwal SG (2022) Five-year ground-based observation trend of PM2. 5 and PM10, and comparison with MERRA-2 data over India. Asian J Atmos Environ 16(3):2022039

3. Anderson JO, Thundiyil JG, Stolbach A (2012) Clearing the air: a review of the effects of particulate matter air pollution on human health. J Med Toxicol 8:166–175
4. Belle JH, Liu Y (2016) Evaluation of aqua modis collection 6 aod parameters for air quality research over the continental united states. Remote Sens 8(10):815
5. Bracaglia M, Volpe G, Colella S, Santoleri R, Braga F, Brando VE (2019) Using overlapping VIIRS scenes to observe short term variations in particulate matter in the coastal environment. Remote Sens Environ 233:111367
6. Brinksma EJ, Pinardi G, Volten H, Braak R, Richter A, Schönhardt A, Levelt PF (2008) The 2005 and 2006 DANDELIONS NO2 and aerosol intercomparison campaigns. J Geophys Res Atmos 113(D16)
7. Collis RTH, Russell PB (2005) Lidar measurement of particles and gases by elastic backscattering and differential absorption. Laser Monitor Atmos 71–151
8. Devaraj S, Tiwari S, Ramaraju HK, Dumka UC, Sateesh M, Parmita P, Shivashankara GP (2019) Spatial and temporal variation of atmospheric particulate matter in Bangalore: a technology-intensive region in India. Arch Environ Contam Toxicol 77:214–222
9. Hao H, Wang K, Zhao C, Wu G, Li J (2024) Visibility-derived aerosol optical depth over global land from 1959 to 2021. Earth Syst Sci Data 16(7):3233–3260
10. He L, Lin A, Chen X, Zhou H, Zhou Z, He P (2019) Assessment of MERRA-2 surface PM2. 5 over the Yangtze River Basin: ground-based verification, spatiotemporal distribution and meteorological dependence. Remote Sens 11(4):460
11. Hendrick F, Müller JF, Clémer K, Wang P, De Mazière M, Fayt C, Van Roozendael M (2014) Four years of ground-based MAX-DOAS observations of HONO and NO 2 in the Beijing area. Atmos Chem Phys 14(2):765–781
12. Hönninger G, Platt U (2002) Observations of BrO and its vertical distribution during surface ozone depletion at alert. Atmos Environ 36(15–16):2481–2489
13. Hu K, Kumar KR, Kang N, Boiyo R, Wu J (2018) Spatiotemporal characteristics of aerosols and their trends over mainland China with the recent Collection 6 MODIS and OMI satellite datasets. Environ Sci Pollut Res 25:6909–6927
14. Huang RJ, Zhang Y, Bozzetti C, Ho KF, Cao JJ, Han Y, Prévôt AS (2014) High secondary aerosol contribution to particulate pollution during haze events in China. Nature 514(7521):218–222
15. Irie H, Kanaya Y, Akimoto H, Iwabuchi H, Shimizu A, Aoki K (2008) First retrieval of tropospheric aerosol profiles using MAX-DOAS and comparison with lidar and sky radiometer measurements. Atmos Chem Phys 8(2):341–350
16. Jolliet O, Antón A, Boulay AM, Cherubini F, Fantke P, Levasseur A, Frischknecht R (2018) Global guidance on environmental life cycle impact assessment indicators: impacts of climate change, fine particulate matter formation, water consumption and land use. Int J Life Cycle Assess 23:2189–2207
17. Kim KH, Kabir E, Kabir S (2015) A review on the human health impact of airborne particulate matter. Environ Int 74:136–143
18. Lewis J, De Young R, Ferrare R, Chu DA (2010) Comparison of summer and winter California central valley aerosol distributions from lidar and MODIS measurements. Atmos Environ 44(35):4510–4520
19. Mahesh B, Rama BV, Spandana B, Sarma MSSRKN, Niranjan K, Sreekanth V (2019) Evaluation of MERRAero PM2. 5 over Indian cities. Adv Space Res 64(2):328–334
20. Mahmud A, Hixson M, Hu J, Zhao Z, Chen SH, Kleeman MJ (2010) Climate impact on airborne particulate matter concentrations in California using seven year analysis periods. Atmos Chem Phys 10(22):11097–11114
21. Malik A, Aggarwal SG (2021) A review on the techniques used and status of equivalent black carbon measurement in two major Asian countries. Asian J Atmos Environ 15(3):2021044
22. McPhetres A, Aggarwal S (2018) An evaluation of MODIS-retrieved aerosol optical depth over AERONET sites in Alaska. Remote Sens 10(9):1384
23. Navinya CD, Vinoj V, Pandey SK (2020) Evaluation of PM2. 5 surface concentrations simulated by NASA's MERRA version 2 aerosol reanalysis over India and its relation to the air quality index. Aerosol Air Qual Res 20(6):1329–1339

24. Pal P, Mukherjee A, Acharya S, Das J (1999) Continuous detection of atmospheric patterns from SODAR signals. Signal Process 74(2):153–168
25. Panicker AS, Athulya S, Ramakrishna K, Beig G, Das S (2024) Particulate matter (PM2. 5) radiative forcing over three different cities in India. Int J Eng Appl Sci Technol 9(1):104–109
26. Pant P, Lal RM, Guttikunda SK, Russell AG, Nagpure AS, Ramaswami A, Peltier RE (2019) Monitoring particulate matter in India: recent trends and future outlook. Air Qual Atmos Health 12:45–58
27. Phalen RF, Cuddihy RG, Fisher GL, Moss OR, Schlesinger RB, Swift DL, Yeh HC (1991) Main features of the proposed NCRP respiratory tract model. Radiat Prot Dosim 38(1–3):179–184
28. Provençal S, Buchard V, da Silva AM, Leduc R, Barrette N, Elhacham E, Wang SH (2017) Evaluation of PM2. 5 surface concentration simulated by Version 1 of the NASA's MERRA aerosol reanalysis over Israel and Taiwan. Aerosol Air Qual Res 17(1):253
29. Randles CA, Da Silva AM, Buchard V, Colarco PR, Darmenov A, Govindaraju R, Flynn CJ (2017) The MERRA-2 aerosol reanalysis, 1980 onward. Part I: system description and data assimilation evaluation. J Clim 30(17):6823–6850
30. Shukla K, Aggarwal SG (2021) Performance check of beta gauge method under high PM2. 5 mass loading and varying meteorological conditions in an urban atmosphere. Atmos Pollut Res 12(11):101215
31. Shukla K, Aggarwal SG (2022) A technical overview on beta-attenuation method for the monitoring of particulate matter in ambient air. Aerosol Air Qual Res 22(12):220195
32. Sigrist MW (ed) (1994) Air monitoring by spectroscopic techniques (Vol. 127. Wiley, New York, pp 163–238
33. Tiwari S, Srivastava AK, Bisht DS, Parmita P, Srivastava MK, Attri SD (2013) Diurnal and seasonal variations of black carbon and PM2. 5 over New Delhi, India: influence of meteorology. Atmos Res 125:50–62
34. Vlemmix T, Hendrick F, Pinardi G, De Smedt I, Fayt C, Hermans C, Van Roozendael M (2015) MAX-DOAS observations of aerosols, formaldehyde and nitrogen dioxide in the Beijing area: comparison of two profile retrieval approaches. Atmos Measur Tech 8(2):941–963
35. Wang Y, Lampel J, Xie P, Beirle S, Li A, Wu D, Wagner T (2017) Ground-based MAX-DOAS observations of tropospheric aerosols, NO 2, SO 2 and HCHO in Wuxi, China, from 2011 to 2014. Atmos Chem Phys 17(3):2189–2215
36. Wei G, Li L, Chen LJ, Zha Y (2022) Impact of fine-mode fraction on the relationship between MODIS AOD and ground-based particulate matter. J Ind Soc Remote Sens 1–9
37. Wittrock F, Oetjen H, Richter A, Fietkau S, Medeke T, Rozanov A, Burrows JP (2004) MAX-DOAS measurements of atmospheric trace gases in Ny-Ålesund-radiative transfer studies and their application. Atmos Chem Phys 4(4):955–966
38. Wu J, Cheng W, Lu H, Shi Y, He Y (2018) The effect of particulate matter on visibility in Hangzhou, China. J Environ Sci Manage 21(1)
39. Zhao H, Che H, Zhang X, Ma Y, Wang Y, Wang H, Wang Y (2013) Characteristics of visibility and particulate matter (PM) in an urban area of Northeast China. Atmos Pollut Res 4(4):427–434

Multi-variable Optimization in Electric Discharge Machining of AISI304 Stainless Steel Using Grey Entropy Weight Method

Sudhir Kumar, Anurag Rai, Arpita Gupta, and Manisha Sharma

Abstract In the present industrial environment, there is fierce competition, which has resulted in a growing demand for highly precise machining of various materials using advanced machining techniques. Electric discharge machining (EDM) is a widely used and extremely efficient advanced machining method that is particularly suitable for working with difficult materials, such as super-alloys, composites, and heat-resistant steels. This study focuses on examining the correlation between the input and output variables of electric discharge machining. Three adjustable machining parameters, specifically current, pulse on time, and voltage, have been selected to determine the material removal rate and surface roughness of AISI304 stainless steel. The experiment is conducted using the Taguchi technique orthogonal array (L27) and the response surface approach is employed to determine the optimal process parameters that minimize surface roughness (SR) and maximize material removal rate (MRR). The efficacy of the present mathematical models has been tested using analysis of variance. This research aims to simultaneously optimize the material removal rate and surface roughness by utilizing the grey entropy weight approach. The experimental results demonstrated that the utilization of the grey entropy methodology significantly improved the output responses in the EDM process compared to the mean method. Furthermore, the grey relational grade was measured to be 0.9641. Hence, the choice of a mechanism for allocating weights to responses and the optimization method itself are pivotal elements in the process

S. Kumar (✉)
Government Polytechnic Uttawar, Palwal, Haryana, India
e-mail: skmeiitdhn@gmail.com

A. Rai
Sharda School of Engineering and Technology, Sharda University, Greater Noida, Uttar Pradesh, India

A. Gupta
Dronacharya Group of Institutions, Greater Noida, Uttar Pradesh, India

M. Sharma
KIET Group of Institutions, Ghaziabad, Uttar Pradesh, India
e-mail: manisha.sharma@kiet.edu

S. Yadav et al. (eds.), *Advances in Metrology*, Lecture Notes in Mechanical Engineering,
https://doi.org/10.1007/978-981-96-6418-4_24

of making decisions for multi-objective optimization. The experimental verification validates the effectiveness of the proposed strategy in enhancing the overall machining performance.

Keywords Electrical discharge machining · Material removal rate · Surface roughness · Analysis of variance · Grey entropy method

1 Introduction

Advanced machining processes are essential for producing intricate components, especially in the aerospace, automotive, and medical sectors [1]. One such technique is electrical discharge machining (EDM), which has recently attracted significant attention due to its capacity to work with hard-to-cut materials and create complex geometries [2]. EDM is a widely utilized manufacturing process that confidently employs controlled electrical discharges to remove material from a workpiece. This process excels in machining challenging, conductive materials, as depicted in Fig. 1, which illustrates the material removal mechanism based on the erosive effect of electric sparks between the workpiece and the electrode [3]. EDM research has made significant advancements over the past decade. Many researchers are working to improve various aspects of the EDM process, including achieving higher material removal rates, enhancing the surface finish of machined components, and ensuring better dimensional accuracy [4]. Researchers worldwide have focussed on exploring ways to optimize the process parameters of the EDM technique. The advancements in EDM have made it possible to fabricate intricate components with mirror-like finishes, modify moulds and dies, and even machine-insulating materials and micro-products.

Table 1 provides perspective for the research effort addressed in this article by presenting relevant studies undertaken by multiple scholars in the last several years and their exceptional outcome(s). The contributors figure out the different factors that need to be changed to make the process work better.

Based on the available literature and the author's expertise, it appears that there is no previous documentation or reporting on the optimization of EDM using specific parameters to achieve desired responses through the integration of the Grey entropy weight (GEW) method for AISI304 stainless steel. Therefore, this current study aims to comprehensively explore and optimize both material removal rate (MRR) and surface roughness (SR) simultaneously using the GEW method. The experimental study follows the Taguchi L27 orthogonal array method, and the responses are evaluated using the response surface methodology for MRR and SR. The study focusses on three main EDM parameters: current, pulse time, and gap voltage, with the output being MRR and SR.

Subsequently, a more thorough investigation was carried out to examine the influence of machine control variables on the measured machining performances. This was achieved by constructing 3D response plots for each specific response variable. In

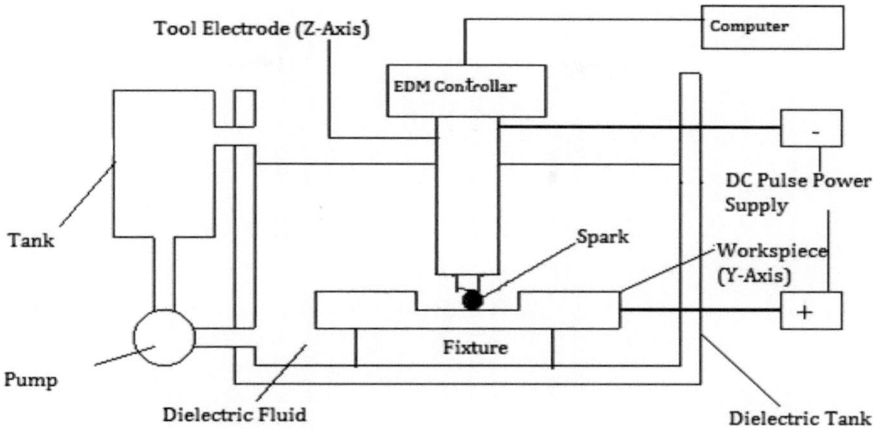

Fig. 1 Schematic diagram of sinking-EDM [5]

order to assess the importance of these factors, the ANOVA technique is utilized, and via this procedure, the most influential model terms are identified. Within the domain of multi-objective optimization, it has been customary for numerous researchers to allocate identical weights to both their models and experiments. Nevertheless, it is essential to acknowledge that when implementing these answers in real-world scenarios, the allocation of weights should correspond to the specific requirements of the industrial sector. The GEW approach is employed for multi-objective optimization, and confirmation trials were done to validate its accuracy. The authors firmly believe that the results of this research will provide useful insights to scholars and industries involved in the field of EDM. Hence, this paper's introduction of novelty and invention is a substantial and valuable contribution to the existing literature.

2 Experimental Set-Up and Measurements

The AISI304 stainless steel has been the focus of an investigation of experimentation. Heavy machinery, automobile, and railway components are some of the many uses for AISI 304 stainless steel. The AISI304 stainless steel is highly resistant to corrosion and oxidation. For this experimental EDM work, the dimensions of the workpiece's material have been chosen as (45 × 20 × 22 mm), using copper tool electrode and EDM oil grade 5-based dielectric medium has been used. Tables 2 and 3 show the characteristics displayed by the electrode material and composition of weight percentage of AISI304 stainless steel. A high-sensitivity weight-measuring device determines the material removal rate (MRR). A surface roughness tester made by Mitutoyo was used to measure the surface roughness (SR).

Table 1 Relevant summarization of recent research literature

S. no.	Author (s) (Year)	Method used	Summarized abstract	Conclusion	Practical implication (s)
1	Naeim et al. (2023)	– Full factorial design of the experiment – MATLAB image processing toolbox for slot characterization	– Wire-EDM of stainless steel 304: surface roughness (SR), material removal rate (MRR) – Process parameters influence MRR and SR significantly	– Process parameters affect MRR and SR significantly – Multi-objective optimization technique needed for high MRR and low roughness	– Optimize process parameters for high MRR and low roughness – Need for multi-objective optimization technique to balance conflicting responses [6]
2	Singh et al. (2022)	– Adaptive network-based fuzzy inference system (ANFIS) for parametric analysis – Grey relational analysis (GRA)-Principal component analysis (PCA) for multi-objective optimization	– Study explores micro-hole machining in super duplex stainless steel – ANFIS and GRA-PCA optimize process parameters for improved results	– Increase in pulse on time (T_{on}) increases MRR – GRA-PCA optimization enhances MRR, reduces Overcut (OC) and Taper angle (TA) significantly	– Improve micro-hole machining in super duplex stainless steel – Enhance MRR, reduce OC and TA [7]
3	Natarajan et al. (2022)	– Taguchi method, Analysis of variance (ANOVA), GRA for optimization of Wire-EDM parameters – Various methods for analysing MRR and SR optimization	– Study on Wire-EDM process parameters optimization for stainless steel – Utilizes Taguchi method, ANOVA, and GRA for best results	– Wire-EDM process parameters significantly impact MRR and surface topography – Taguchi method and GRA are effective optimization methods	– Wire-EDM optimizes process parameters for better material removal rate – Taguchi and GRA methods enhance machining efficiency and accuracy [8]

(continued)

Table 1 (continued)

S. no.	Author (s) (Year)	Method used	Summarized abstract	Conclusion	Practical implication (s)
4	Hema and Aparna (2021)	– Taguchi method based grey relational analysis – L9 Orthogonal array for parameter combinations	– Optimal Wire-EDM parameters for Monel 400 and AISI 316 weldment – T_{on} crucial; 150 μs, pulse off time (T_{off})40 μs, 14 kg-f tension optimal	– T_{on} has a significant effect on output parameters – Optimal process parameters: T_{on} 150 μs, T_{off} 40 μs, wire tension 14 kg-f	– Optimal parameters for efficient Wire-EDM on Monel 400 and AISI 316 – Improved MRR, SR, and kerf width control [9]
5	Camposeco-Negrete (2021)	– Orthogonal array, signal-to-noise ratio, means graphs, ANOVA – Multi-objective optimization for sustainable machining approach	– Study optimized cutting parameters in wire-EDM for sustainability and productivity – Multi-objective optimization reduced electricity cost and CO_2 emissions in machining	– T_{on} is most significant factor in wire-EDM process – Sustainable approach reduces electricity cost and CO_2 emissions	– Optimize wire-EDM parameters for energy efficiency and quality improvement – Achieve sustainable machining with reduced electricity cost and CO_2 emissions [10]
6	Zhou et al. (2020)	– Variance of machining states used as an index – Adaptive gap servo-voltage control system implemented	– EDM of Inconel 718 faces challenges due to material properties – Proposed adaptive gap servo-voltage control system for stable machining	– Proposed variance of machining states as an index – Constructed adaptive gap servo-voltage control system for EDM stability	– Improved stability and efficiency in machining Inconel 718 using EDM – Adaptive gap servo-voltage control system enhances machining of difficult materials [11]

(continued)

Table 1 (continued)

S. no.	Author (s) (Year)	Method used	Summarized abstract	Conclusion	Practical implication (s)
7	Ahmed et al. (2022)	– Hybrid EDM ECM process – Addition of salt additives for controlled removal mechanisms	– Hybrid EDM ECM process for zero recast surfaces proposed – Salt additives control removal mechanisms, producing tailored surfaces	– Hybrid EDM ECM process for zero recast surfaces proposed – Salt additives control removal mechanisms, producing tailored surfaces	– Hybrid EDM ECM process for zero recast surfaces – Complete removal of defective recast layer in single step [12]
8	Escobar et al. (2020)	– Proposed model for plasma channel formation – Evaluation using experimental data and existing models	– EDM process removes metal using electric spark erosion – New model proposed for plasma channel formation in EDM process	– Model type F 1 precise for medium/high energy, not low – Proposed model provides precision in heat flow and discharge energy	– Model predicts plasma channel formation in EDM process – Precision varies with energy ranges, better for medium to high [13]

(continued)

Table 1 (continued)

S. no.	Author (s) (Year)	Method used	Summarized abstract	Conclusion	Practical implication (s)
9	Kishi and Yan (2020)	– Electrical discharge machining (EDM) – Electrochemical machining (ECM)	– Proposed hybrid process of EDM and ECM using same tool electrode – Surface defects removed and surface roughness improved with ECM finishing process	– ECM after EDM improves SR and removes defects – Hybrid process enhances productivity and surface integrity for microscale products	– Improved surface integrity and productivity in microscale product fabrication – Prevents tool alignment errors by using same machine and electrode [14]
10	Wang et al. (2023)	– Investigated effect of discharge current, electrode diameter, depth of cut – Developed electrode end-face design system to improve geometric accuracy	– Novel electrode design to enhance geometric accuracy in EDM process – Proposed method improves accuracy by around 22–37%	– Discharge current most significant, followed by depth, then diameter – Proposed method enhances geometric accuracy by 22–37%	– Proposed electrode design improves workpiece corner accuracy by 22–37% – System compensates electrode wear to enhance geometric accuracy [15]

Table 2 Properties of tool electrode material [5]

Electrode material	Density (gm/cc)	Specific heat (J/kg.K)	Thermal conductivity (W/mK)	Electrode resistivity (μ-ohms)	Hardness (BHN)
Copper	8.9	386	399	1.69	48

Table 3 AISI304 stainless steel composition in weight %

Chemical composition of AISI420					
Cr	*Ni*	*Mn*	*Si*	*C*	*P*
19.6	9.4	2.2	0.73	0.048	0.045
S	*N*	*Fe*			
0.028	0.11	Rest			

Table 4 Machining parameters and their levels for design of experiments [5]

Machine control variables	Unit	Code	Range (as specified by machine manufacturer)	Levels and values		
				1	2	3
Gap voltage	Volt (V)	V	15–150	55	65	75
Current	Ampere (A)	I_P	0–10	4	6	8
Pulse on time	μsec	T_{on}	0.5–2000	50	100	150

The study examined the impact of the input factors on the replies by considering them at three different levels. The process parameters, along with their corresponding levels, are presented in Table 4. In order to reduce the margin of error, every experiment has been conducted on three separate occasions, and the resulting average value is taken into account. Following each experiment, thorough cleaning of both the workpiece and electrode is conducted, after which the weight of both components is measured. Carbon buildup on the electrode has been overlooked by authors.

3 Methodology

The experiments were meticulously planned and executed following the principles of DOE. DOE is a methodical approach aimed at maximizing information acquisition while minimizing the number of experiments required. Numerous experimental design options fall under the umbrella of DOE, with one of the most widely adopted approaches is response surface methodology [16]. RSM, a mathematical modelling technique, serves the purpose of establishing relationships between dependent and independent variables [17]. The entire experimental plan, consisting of 27 distinct experiments, was meticulously designed using Design Expert v13 statistical analysis software. Table 5 provides a detailed breakdown of 27 experiment trials, including consideration of the three parameters at three different levels.

The chosen-response MRR falls into the "Larger-the-better" (LTB) classification, while SR belongs to the "smaller-the-better" (STB) category. The term "Grey" refers to information that is characterized by its inadequacy, ambiguity, and lack of completeness. To maintain the precision and credibility of research, it is imperative to gather comprehensive data that meets well-defined and quantifiable criteria, a standard upheld by conventional statistical methods. Addressing the challenge posed

Table 5 Experimental results of measured machining performances (MMP)

Exp. no	Gap voltage (V)	Current (I_p)	Pulse on time (T_{on})	MRR (g/min)	SR (μm)
1	55	4	50	3.70	5.67
2	55	4	50	4.83	3.58
3	55	4	50	5.62	2.41
4	65	4	100	3.30	8.99
5	65	4	100	3.60	7.78
6	65	4	100	3.57	7.48
7	75	4	150	3.70	10.49
8	75	4	150	3.59	9.72
9	75	4	150	3.14	9.86
10	65	6	50	4.42	5.61
11	65	6	50	5.19	4.53
12	65	6	50	5.62	4.38
13	75	6	100	5.44	6.42
14	75	6	100	5.39	6.22
15	75	6	100	5.00	6.94
16	55	6	150	6.54	6.67
17	55	6	150	6.08	6.91
18	55	6	150	5.28	8.06
19	75	8	50	5.22	6.68
20	75	8	50	5.64	6.61
21	75	8	50	5.72	7.46
22	55	8	100	7.67	3.98
23	55	8	100	7.25	5.78
24	55	8	100	6.51	7.51
25	65	8	150	9.48	3.97
26	65	8	150	8.66	5.21
27	65	8	150	7.50	7.38

by ambiguous and multifaceted information, grey theory offers a suitable explanation. GRA, built upon grey theory, specializes in identifying and exploring uncertain relationships between a pivotal factor and all other pertinent elements within a designated system [18]. Considering the aim of this paper, the focus lies on maximizing the MRR. Consequently, for the LTB criterion for these quality attributes, and the normalized results can be succinctly represented as Eq. (1):

$$K'_{ij} = \frac{L'_{ij} - min_j L'_{ij}}{max_i L'_{ij} - min_j L'_{ij}} \tag{1}$$

Further, SR is minimized, thus the STB is used, as expressed in Eq. (2):

$$K'_{ij} = \frac{max_j K' - L'_{ij}}{max_i L'_{ij} - min_j L'_{ij}} \tag{2}$$

Data normalization is succeeded by the computation of Grey Relational Coefficient (GRC), revealing the connection between ideal and actual normalized experimental outcomes. The application of the GRC is demonstrated through Eq. 3:

$$\xi'_{ij} = \frac{min_i min_j \left|K_i^o - K'_{ij}\right| + \zeta max_i max_j \left|K_i^o - K'_{ij}\right|}{\left|K_i^o - K'_{ij}\right| + \zeta max_i max_j \left|K_i^o - K'_{ij}\right|} \tag{3}$$

The Grey Relational Grade (GRG) is a performance index used to evaluate the multiple performance characteristics of a system. In the next step, the Grey Relational Grade (GRG) is determined through the equalization of GRC:

$$\gamma'_{ij} = \frac{1}{M} \sum\nolimits_{i=1}^{M} \xi_{ij} \tag{4}$$

In specific practical scenarios, the weights of MMPs vary, including weights derived from grey entropy considerations. The desired weights for replies are calculated using the GEW approach, developed by Shannon and Weaver in 1947 and Zeleny in 1982 [19]. The GEW method utilizes probability theory to quantify ambiguous information in order to evaluate the importance of each MMP. The decision-maker's preferences are not taken into account in this assessment [20]. The GEW method presupposes that higher quality weight predictor data is more important. This method defines objectives using a decision matrix, then estimates, normalizes, assesses chances, computes, determines levels of a divergence, and derives entropy weight [21]. The data collected for the experiment is normalized using linear normalization to account for MMP unit variability. Equation (5) calculates favourable MMP (MRR), while Eq. (6) calculates unfavourable MMPs (SR) and notice that normalized decision matrix (NDM_{ij}) € [0,1]:

$$NDM_{ij} = \frac{L'_{ij}}{max_i L'_{ij}} \tag{5}$$

$$NDM_{ij} = \frac{\min_i K\prime_{ij}}{K\prime_{ij}} \tag{6}$$

The probability of the response (Pr_{ij}) is calculated by Eqs. (7) and (8) is utilized to compute the Entropy (En_j):

$$Pr_{ij} = \frac{NDM_{ij}}{\sum_{i=1}^{n} NDM_{ij}} \tag{7}$$

$$En_j = -\frac{\sum_{i=1}^{n} Pr_{ij} \log_e(Pr_{ij})}{\log_e(n)} \tag{8}$$

Equation (9) is used to calculate the degrees of divergence (Div_j), and Eq. (10) obtains the entropy weight (E_w). In such cases, Eq. (4) is modified to Eq. (11):

$$Div_j = \left|1 - En_j\right| \tag{9}$$

$$Ew_j = \frac{Div_j}{\sum_{j=1}^{m} Div_j} \tag{10}$$

$$\gamma\prime_{ij} = \frac{1}{M} \sum\nolimits_{i=1}^{M} Ew_j \xi_{ij} \tag{11}$$

Better multi-response characteristics are indicated by a greater GRG. All letters and symbols in Eqs. 1–11 have standard interpretations. The complete experimental planning encompasses several stages, as illustrated in Fig. 2.

4 Results and Discussion

Effect of process parameters on responses

Figure 3 shows the 3D response plot for MRR and SR. The application of higher voltages typically results in increased MRR, especially at elevated feed rates, which are attributable to the augmented discharge gap and intensified sparking. Excessive voltage levels may lead to diminished MRR due to reduced spark intensity within

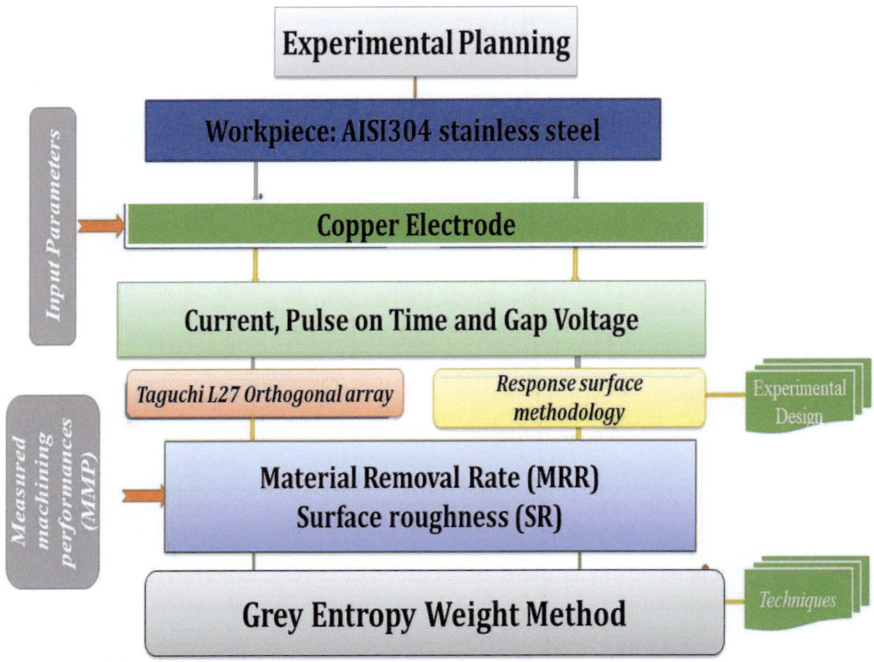

Fig. 2 Experimental planning of current research work

the dielectric fluid. Increasing the current results in more powerful sparks, leading to higher energy density and superior material removal. However, it's essential to be mindful of the limitations of high currents, as they can adversely affect other aspects of the machining process. Optimizing the T_{on} in EDM process for AISI 304 stainless steel is crucial for maximizing the MRR. An initial increase in Ton duration leads to a rise in MRR until it reaches an optimal value. However, it's essential to be mindful that beyond this point, the MRR may decrease due to unstable discharges and reduced process efficiency.

Increasing the voltage has a positive effect on reducing the SR of the machined part until it reaches an optimal voltage level. Higher voltages create a more stable machining environment by widening the discharge gap. However, avoiding excessively high voltages is essential as they can negatively impact surface quality. SR shows an initial increase with T_{on} but declines after reaching an optimal value. This decline is due to the onset of unstable discharges and reduced process efficiency at excessively high T_{on} values. Boosting the I_p results in greater energy density and increased crater formation irregularities, reducing surface finish.

Multi-response Optimization Based on Grey Entropy Weight (GEW) Method

In GRA, a uniform allocation of equal weights to all response variables can introduce an element of uncertainty into the decision-making process. To mitigate this, the

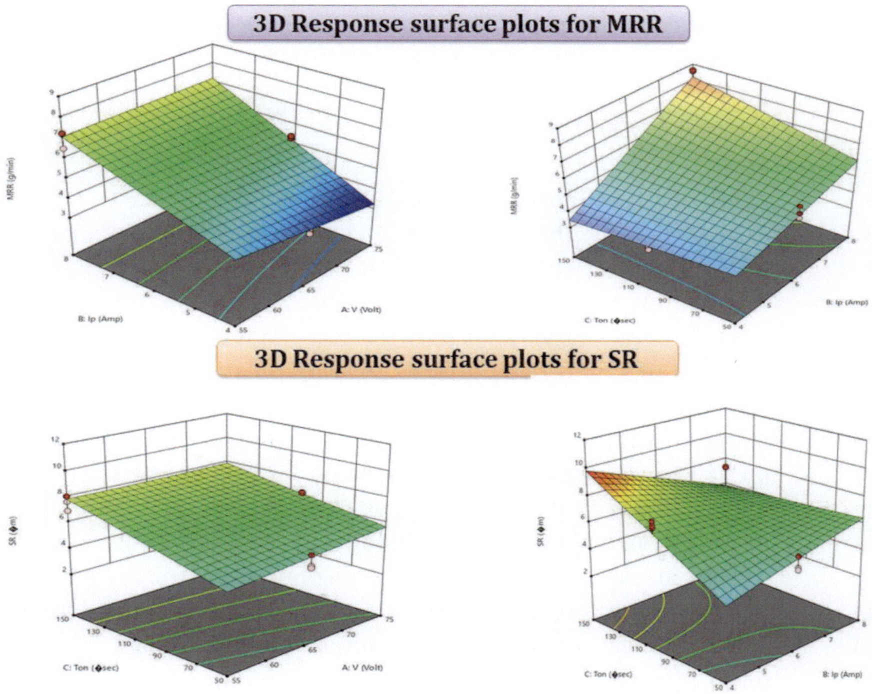

Fig. 3. 3D response surface plots for MRR and SR

GEW method is adopted, offering a means to ascertain the relative weights of MMPs. In this research endeavour, the aim is to conduct a comparative analysis of multi-objective optimization achieved through the utilization of the GEW method. Initially, the experimental data, as outlined in Table 5 and encompassing measurements of MMPs, is subjected to normalization through the application of Eqs. (1) and (2). The computation of the GRC for each individual MMP is accomplished through the utilization of Eq. (3). Subsequently, the GRG (mean method) is determined utilizing Eq. (4) and illustrated in Table 6. For the GEW method, Eqs. (5) and (6), facilitating linear normalization of the experimental data to account for the diverse units associated with the MMPs. Following this, the Pr_{ij} is determined using Eq. (7). Subsequently, En_j is computed as per Eq. (8). Div_j are derived through the application of Eq. (9), while Eq. (10) is employed to obtain the E_w. Employing the weights derived through the GEW method, the GRGs are computed using Eq. (11). Table 7 depicts the ranking of these GRGs as determined by the GEW method. Notably, in Tables 6 and 7, the highest GRG value achieved through the mean method corresponds to sample number 25, while the highest GRG value computed via the GEW method corresponds to sample number 24. Furthermore, as indicated in Fig. 4, the means of GRG values obtained through the mean method and GEW methods reveal identical optimal conditions.

Table 6 Normalized, deviation sequence, GRC and GRG (mean method)

Exp. no	Normalized values		Deviation sequence		GRC		GRG	Rank
	MRR	SR	MRR	SR	MRR	SR		
1	0.0883	0.5965	0.9117	0.4035	0.3542	0.5534	0.4538	18
2	0.2666	0.8552	0.7334	0.1448	0.4054	0.7754	0.5904	5
3	0.3912	1.0000	0.6088	0.0000	0.4509	1.0000	0.7255	2
4	0.0252	0.1856	0.9748	0.8144	0.3390	0.3804	0.3597	24
5	0.0726	0.3354	0.9274	0.6646	0.3503	0.4293	0.3898	23
6	0.0678	0.3725	0.9322	0.6275	0.3491	0.4435	0.3963	22
7	0.0883	0.0000	0.9117	1.0000	0.3542	0.3333	0.3438	26
8	0.0710	0.0953	0.9290	0.9047	0.3499	0.3559	0.3529	25
9	0.0000	0.0780	1.0000	0.9220	0.3333	0.3516	0.3425	27
10	0.2019	0.6040	0.7981	0.3960	0.3852	0.5580	0.4716	14
11	0.3233	0.7376	0.6767	0.2624	0.4249	0.6558	0.5404	8
12	0.3912	0.7562	0.6088	0.2438	0.4509	0.6722	0.5616	7
13	0.3628	0.5037	0.6372	0.4963	0.4397	0.5019	0.4708	16
14	0.3549	0.5285	0.6451	0.4715	0.4366	0.5146	0.4756	13
15	0.2934	0.4394	0.7066	0.5606	0.4144	0.4714	0.4429	20
16	0.5363	0.4728	0.4637	0.5272	0.5188	0.4867	0.5028	10
17	0.4637	0.4431	0.5363	0.5569	0.4825	0.4731	0.4778	12
18	0.3375	0.3007	0.6625	0.6993	0.4301	0.4169	0.4235	21
19	0.3281	0.4715	0.6719	0.5285	0.4266	0.4862	0.4564	17
20	0.3943	0.4802	0.6057	0.5198	0.4522	0.4903	0.4713	15
21	0.4069	0.3750	0.5931	0.6250	0.4574	0.4444	0.4509	19
22	0.7145	0.8057	0.2855	0.1943	0.6365	0.7201	0.6783	4
23	0.6483	0.5829	0.3517	0.4171	0.5870	0.5452	0.5661	6
24	0.5315	0.3688	0.4685	0.6312	0.5163	0.4420	0.4791	11
25	1.0000	0.8069	0.0000	0.1931	1.0000	0.7214	**0.8607**	**1**
26	0.8707	0.6535	0.1293	0.3465	0.7945	0.5906	0.6926	3
27	0.6877	0.3849	0.3123	0.6151	0.6155	0.4484	0.5320	9

The ANOVA technique serves as an effective means to assess the relative influence of machining variables on responses. In this research paper, the impact of MCVs on GRG (mean method) and GRG (GEW method), determined through the p-value at a 95% confidence level. Notably, all three MCVs exhibited a substantial effect on GRG (mean method) and GRG (GEW method), as evidenced by their model term p-values. Furthermore, the lack-of-fit term was deemed insignificant, affirming the adequacy of our proposed model for predicting the output variables. For GRG (mean method), the coefficient of determination (R-sq.) value, standing at 94.61%, signifies that the control factors account for a substantial 94.61% of the variation

Table 7 Probability criterion, entropy, divergence entropy weight and GRG (GEW method)

Exp. no	P_{ij}		lnPij		Pij*lnPij		GRG (GEW) method	GRG (GEW) rank
	MRR	SR	MRR	SR	MRR	SR		
1	0.0883	0.5965	−3.6866	−3.4370	−0.3256	−2.0503	0.3799	21
2	0.2666	0.8552	−3.4201	−3.8968	−0.9117	−3.3326	0.4530	14
3	0.3912	1.0000	−3.2686	−4.2926	−1.2786	−4.2926	0.5217	6
4	0.0252	0.1856	−3.8010	−2.9761	−0.0959	−0.5525	0.3444	26
5	0.0726	0.3354	−3.7140	−3.1206	−0.2695	−1.0466	0.3605	23
6	0.0678	0.3725	−3.7223	−3.1600	−0.2525	−1.1772	0.3613	22
7	0.0883	0.0000	−3.6866	−2.8218	−0.3256	0.0000	0.3515	24
8	0.0710	0.0953	−3.7168	−2.8980	−0.2638	−0.2762	0.3507	25
9	0.0000	0.0780	−3.8507	−2.8837	0.0000	−0.2248	0.3357	27
10	0.2019	0.6040	−3.5088	−3.4476	−0.7084	−2.0822	0.4074	20
11	0.3233	0.7376	−3.3482	−3.6615	−1.0826	−2.7008	0.4547	13
12	0.3912	0.7562	−3.2686	−3.6951	−1.2786	−2.7942	0.4794	10
13	0.3628	0.5037	−3.3011	−3.3128	−1.1976	−1.6687	0.4477	15
14	0.3549	0.5285	−3.3104	−3.3444	−1.1748	−1.7674	0.4467	16
15	0.2934	0.4394	−3.3855	−3.2349	−0.9932	−1.4213	0.4217	19
16	0.5363	0.4728	−3.1170	−3.2746	−1.6716	−1.5481	0.5147	7
17	0.4637	0.4431	−3.1899	−3.2392	−1.4792	−1.4352	0.4813	9
18	0.3375	0.3007	−3.3310	−3.0853	−1.1243	−0.9279	0.4284	18
19	0.3281	0.4715	−3.3424	−3.2731	−1.0966	−1.5434	0.4343	17
20	0.3943	0.4802	−3.2650	−3.2836	−1.2875	−1.5768	0.4571	11
21	0.4069	0.3750	−3.2509	−3.1626	−1.3229	−1.1860	0.4558	12
22	0.7145	0.8057	−2.9576	−3.7909	−2.1132	−3.0543	0.6473	3
23	0.6483	0.5829	−3.0139	−3.4178	−1.9538	−1.9923	0.5816	5
24	0.5315	0.3688	−3.1216	−3.1560	−1.6593	−1.1640	0.5067	8
26	1.0000	0.8069	−2.7457	−3.7934	−2.7457	−3.0610	0.9641	1
27	0.8707	0.6535	−2.8362	−3.5216	−2.4694	−2.3012	0.7682	2

in GRG (mean method), with only a minute 5.39% of the total variation eluding description within the confines of our proposed model. Additionally, the adjusted coefficient of determination (Adj. R-Sq)' stands at 93.56%.

As the deviation in between R-sq. and Adj. R-sq. lies within an acceptable range, the model suggested can accurately be considered appropriate and effective. The model's effectiveness to forecast prospective MRR outcomes is shown by the significant level of concurrence displayed in the R-sq values. Furthermore, the standard deviation, quantified at 0.0135 for GRG (mean method), reflects a limit of 0.0135 standard deviations from the mean. Similarly, for GRG (GEW method), the R-sq value, standing at 98.46%, signifies that the control factors account for a substantial

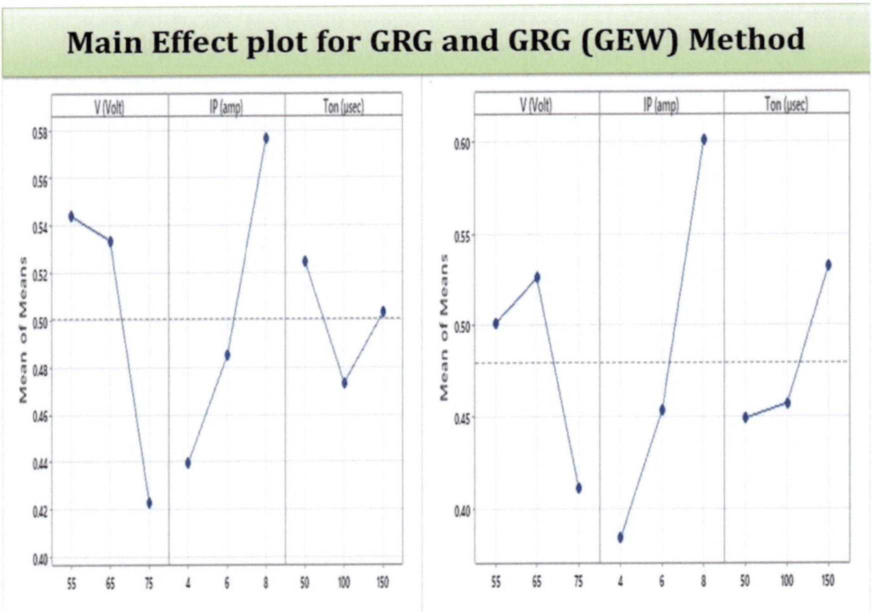

Fig. 4 Main effect plot for GRG and GRG (GEW) method

98.46% of the variation in GRG (GEW method), with only a minute 1.54% of the total variation eluding description within the confines of our proposed model. Additionally, the "Adj. R-Sq." stands at 97.30% and the standard deviation, quantified at 0.0203 for GRG (GEW method). ANOVA for GRG (mean method) and GRG (GEW method) shows that factor I_p is the most impactful MCV. The order of significance based on ANOVA is as follows: $I_p > V > T_{on}$, respectively. In summary, the ANOVA results closely align with the GEW method outcomes.

Confirmation Experiment

Experiment was carried out to confirm the optimal values obtained by the GRG (mean method) and GRG (GEW method) for assessing and verifying the quality response improved by EDM on AISI304 stainless steel. The probable value of GRG, for a given set of parameters, employing the expression shown in Eq. (12):

$$\hat{a} = a_m + \sum_{i=1}^{p} \overline{a} - a_m \tag{12}$$

In Eq. 12, the $GRG_{\text{estimated optimal}}$ machining parameter is calculated, where $GRG_{\text{total mean}}$ represents the GRG_{mean} at the optimum level, and p describes the total quantity of MCVs that impact the measured machining performance. The outcomes of the experiments executed to verify the results, confirming the improvement in multiple performance characteristics through the GEW method. The experimental verification confirms the efficacy of the suggested method in improving the overall machining performance.

5 Conclusions

The effect of process parameters on the EDM process for the AISI304 stainless steel material has been assessed. The results obtained from the experiments exploring process parameters and their impact on MRR and SR are succinctly summarized below:

- The higher voltages typically result in increased MRR, especially at elevated feed rates, which are attributable to the augmented discharge gap and intensified sparking.
- SR shows an initial increase with T_{on} but declines after reaching an optimal value. This decline is due to the onset of unstable discharges and reduced process efficiency at excessively high T_{on} values.
- For GRG (mean method), the coefficient of determination (R-sq.) value stands at 94.61% and the adjusted coefficient of determination (Adj. R-Sq) stands at 93.56%. Similarly, for GRG (GEW method), the R-sq value stands at 98.46% and the "Adj. R-Sq." stands at 97.30%.
- ANOVA for GRG (mean method) and GRG (GEW method) shows that factor I_p as the most impactful MCV. The order of significance based on ANOVA is as follows: $I_p > V > T_{on}$, respectively.
- This methodology is cost-effective, time-saving and efficient, and accurate in optimizing machining parameters.

References

1. Gnanavel C et al (2017) Restructured review on electrical discharge machining—a state of the art, vol 183. IOP Publishing, pp 012015–012015
2. Gohil and Puri (2016) Turning by electrical discharge machining: a review, vol 231(2). SAGE Publishing, pp 195–208
3. Mandal, Dixit (2014) State of art in wire electrical discharge machining process and performance, vol 16(1). Inderscience Publishers, pp 1–1
4. Equbal, Sood (2014) Electrical discharge machining: an overview on various areas of research. Manuf Ind Eng 13:1–2
5. Kumar S et al (2020) Optimization of process variables in electric discharge machining (EDM) using Taguchi methodology. Indian J Eng Mater Sci 27:819–825
6. Naeim N et al (2023) Experimental investigation of surface roughness and material removal rate in wire EDM of stainless steel 304. Materials 16(3):1022
7. Singh DP et al (2022) Parametric analysis through ANFIS modelling and optimization of micro-hole machining in super duplex stainless steel by die-sinking EDM. Adv Mater Process Technol 9(4):1885–1902

8. Natarajan K et al (2022) Study on optimization of WEDM process parameters on stainless steel. J Nanomater 2022:1–7
9. Hema, Aparna (2021) Process parameter optimization of wire EDM on weldment of Monel 400 and AISI 316 grade steel. IOP Conf Ser Mater Sci Eng 1132(1):012027
10. Camposeco-Negrete C (2021) Analysis and optimization of sustainable machining of AISI O1 tool steel by the wire-EDM process. Adv Manuf 9:304–317
11. Zhou S et al (2020) Electrical discharge machining Inconel 718 with adaptively regulating gap servo-voltage. Int J Adv Manuf Technol 109:2575–2585
12. Ahmed S et al (2022) Electrolytic-dielectrics: a route to zero recast electrical discharge machining. Int J Mach Tools Manuf 181:103941
13. Escobar AM et al (2020) Simplified plasma channel formation model for the electrical discharge machining process. Int J Adv Manuf Technol 106:143–153
14. Kishi R, Yan J (2020) Electrical discharge/electrochemical hybrid machining based on the same machine and tool electrode. J Micro Nano-Manuf 8(1):010906
15. Wang SM et al (2023) A novel electrode front-end face design to improve geometric accuracy in electrical discharge machining process. Metals 13(6):1122
16. Box GEP, Wilson KB (1951) On the experimental attainment of optimum conditions. J R Stat Soc 13(1):1–45
17. Myers, Montgomery (1995) Response surface methodology, process and product optimization using designed experiments. Wiley, New York
18. Lin ZC, Ho CY (2003) Analysis and application of Grey relation and ANOVA in chemical–mechanical polishing process parameters. Int J Adv Manuf Technol 21(1):10–14
19. Shannon CE (1948) A mathematical theory of communication. Bell Syst Tech J 27(3):379–423
20. Kumar R, Bilga PS, Singh S (2017) Multi objective optimization using different methods of assigning weights to energy consumption responses, surface roughness and material removal rate during rough turning operation. J Clean Prod 164:45–57
21. Rao R, Yadava VJO, Technology L (2009) Multi-objective optimization of Nd: YAG laser cutting of thin superalloy sheet using grey relational analysis with entropy measurement. Opt Laser Technol 41(8):922–930

Development and Validation of an Automated 50 KN Dead Weight Force Measurement System and Dissemination of Uncertainty of Measurement

Vishal R. Balwan, Raju Tambad, and Ravindra Ghorpade

Abstract Force measurement plays an important role in a wide range of industrial applications, from electronics assembly to pharmaceutical tablet pressing, metal forging, vehicle manufacture, and precision testing. The investigation of the automation of a 50 KN deadweight force calibration machine, critical equipment at the FIE Research Institute, is done in this research. This machine, which was initially controlled manually, presented difficulties in terms of time consumption and operator fatigue. The paper explains how automation was used successfully to streamline the process, optimize productivity, and minimize operator fatigue resulting in a 30% reduction in operational time. Additionally, automation has made machine access more democratic, enabling semi-skilled employees to use it effectively. To validate the machine, standard load cells were tested on it and their output values such as repeatability, uncertainty, and reproducibility were compared with NMI readings.

Keywords Force · Load cell · Automation · FRI · NMI

1 Introduction

Force calibration is crucial when testing materials used to make equipment, machinery, and gadgets. Understanding the characteristics and durability of materials is essential for the safety of both consumers and industrial processes because all metals and other materials can expand and contract when in use [1, 2]. Material properties are important in product design because they affect how long a thing will last. Understanding a material's traits helps determine if it's strong enough to handle the forces it will face [3].

V. R. Balwan (✉) · R. Tambad · R. Ghorpade
FIE Research Institute, Ichalkaranji, Maharashtra, India
e-mail: vrbalwan@gmail.com

S. Yadav et al. (eds.), *Advances in Metrology*, Lecture Notes in Mechanical Engineering,
https://doi.org/10.1007/978-981-96-6418-4_25

1.1 History of Force Calibration

Metal failures led to catastrophic disasters that resulted in property devastation and fatalities in the late nineteenth and early twentieth centuries. At the time, testing procedures were neither regulated nor sufficient to guarantee equipment stability. To avoid further occurrences of this kind, material testing was developed as a result [4, 5]. The first testing tool utilized was a Brinell tester, which tested materials by exerting force against them using a 10 mm metal ball. The ball was propelled by a pneumatic piston, but the drive's variable force produced unreliable and inconsistent results. To remedy this, more accurate measurement tools were created in cooperation with the National Bureau of Standards of the United States. The most precise and trustworthy results are produced by today's force calibration equipment because of the century-long advancement of force calibration techniques [6].

1.2 Force Calibration Measurements

Sir Isaac Newton's renowned saying "Force controls motion" emphasizes how crucial force control is for achieving control over motion. The definition of force in physics is defined as the sum of mass and acceleration. The use of parameters and baselines to assess the accuracy of instruments is necessary for the accurate measurement of force [7].

Force is a kind of mechanical energy that can be either a push or a pull. The amount of force being applied to an object, substance, or device is determined by the data generated by force calibration equipment. Measurements of force are used to assess how much tension, compression, or impact an object can withstand because force is applied to everything. The instrument being used, which needs to be correctly aligned, level, square, sturdy, and with minimum torsion, is the key to successful calibration. The data gathered will be wrong and erroneous if these conditions are not optimum [7, 8].

Although measuring force on items may seem like a simple idea, doing so accurately and reliably needs a lot of technological and scientific know-how. Accuracy and dependability are critical in the case of force calibration since they are necessary for security and to prevent errors. To set parameters for their products, manufacturers and producers rely on force data. Sales information and product manuals can contain this information. For safety reasons, force calibration is required [8].

1.3 Types of Force Calibration

Measuring and cataloguing force is crucial for ensuring the proper performance of a device or material. Over the almost 100 years of force calibration, manufacturers have developed and perfected techniques and methods to precisely fit the needs of any object or material.

The first type is tensile force calibration. Tension calibration is the process of elongating an object and is the opposite of compression. Secondly the compressive force calibration. It involves compressing or squeezing something, which is typically measured as a pushing or squeezing force. For accurate calibration, everything must be properly aligned with the force being applied as straight and true as possible, and free from any unconventional or angular forces. Another is the hardness calibration [9]. Hardness is defined as the resistance of a material to indentation when a fixed force or load is applied to it. The smaller the indentation made on the material, the harder it is. Several methods are used to test a material's hardness, with the indentation's depth or area determining its hardness value. One more is the impact force calibration. The purpose of impact testing is to assess how well a material can withstand sudden impact. This technique involves subjecting the material to sudden stress, such as bending, tension, or torsion, to determine its toughness and brittleness. By measuring how the material behaves after being exposed to sudden stress, an impact test can determine its characteristics in terms of resilience and susceptibility to cracking or fracturing [9, 10].

2 Materials and Methods

The force calibration machine is a specialized equipment created for accurate force measurement. It was created by the FIE Research Institute and is shown in Fig. 1. This device is made up of several carefully calibrated weights that have been precisely made to meet the highest standards of accuracy and reliability. There are two unique portions of weights in this machine: the upper part, which consists of ten smaller weights, each with a force rating of 250 Newtons (N), and the lower section, which consists of ten larger weights, each with a force rating of 5 kilonewtons (KN).

With the ability to choose weights ranging from 250 N to an astonishing 50 KN, in increments of 250 N, this design enables a wide variety of load selections, meeting a variety of force-measuring applications. Three platforms are attached to the machine's upper section. While the central platform is intended to be flexible, the lower and higher platforms stay fixed in their positions to provide stability during testing. The testing conditions may be precisely controlled thanks to this novel design element.

Because the weights are only attached to the middle platform, depending on the testing needs, a given load can produce either compressive forces between the lower and middle platforms or tensile forces between the upper and middle platforms. Due to its adaptability, the machine can replicate a variety of force settings, making it ideal for calibrating tasks in a variety of applications. A load cell can be easily loaded between the platforms for calibration purposes. This load cell serves as a critical component in the calibration process, as it accurately measures and records the forces applied during testing, facilitating precise calibration, and ensuring the

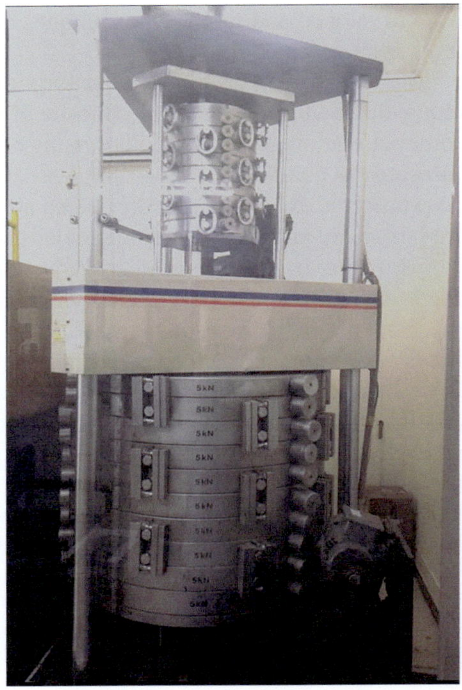

Fig. 1 Deadweight force calibration machine

machine's consistent and reliable performance. A particular procedure is followed to ensure accurate measurements and calibration to conduct load cell testing. Here's a step-by-step overview of the process:

- The first step involves securely attaching the load cell between the machine's platforms, and aligning it according to the specific testing requirements, whether for tensile or compression force measurements.
- Once the load cell is in place, operators have the flexibility to manually select the desired loads based on the testing parameters. This step allows for precise control over the magnitude of force to be applied.
- With the selected loads in hand, they are carefully applied to the middle platform of the machine. As the force is exerted, the load cell responds by registering the exact magnitude of the applied force.
- The output from the load cell is then meticulously observed and recorded. This data can be conveniently accessed through a Digital Read Out (DRO) display or a dedicated load indicator, both of which are integral components of the testing equipment.
- In the event of any discrepancies or errors in the measured data, they are promptly noted and addressed. Corrections are made to ensure the accuracy and reliability of the load cell's performance.

2.1 Automation of Operation

Automation is a key advancement in the operation of the dead weight force calibration machine, streamlining the process of loading, unloading, recording, and auto-calibration. The contemporary version of this machine incorporates various mechanisms to enhance efficiency and accuracy. Operators are provided with a user-friendly interface to input the desired weight value into a controller. The controller, in turn, engages a programmed motor responsible for selecting the appropriate weight. This motor is connected to a screw mechanism that facilitates the smooth loading or unloading of the selected weight.

The automation process is overseen by a Programmable Logic Controller (PLC) and proxy limit switches, ensuring precise and controlled movements. Additionally, a touchscreen panel is integrated to allow for the easy selection of weights.

The jack system used in this setup as shown in Fig. 2 features a 60 mm square-threaded screw and a cast iron nut, meticulously positioned within two ball bearings and a thrust bearing. A sprocket, adhering to a 12.5 mm chain specification, is attached to the nut and drives its movement, raising or lowering the screw. The weights can be lifted or lowered as per loading or unloading with the help of a screw jack assembly.

This entire jack assembly is securely mounted on the base of the machine. The weights themselves rest on a plate connected to the screw.

This automation system is implemented for the 250 N and 5 KN weight stacks. Various load cells and their associated indicators are currently being evaluated for integration with a computer. This data input is recorded and stored for subsequent calculations and certificate generation. The automation also involves the utilization of servo gear motors, each fitted onto a bracket corresponding to a particular

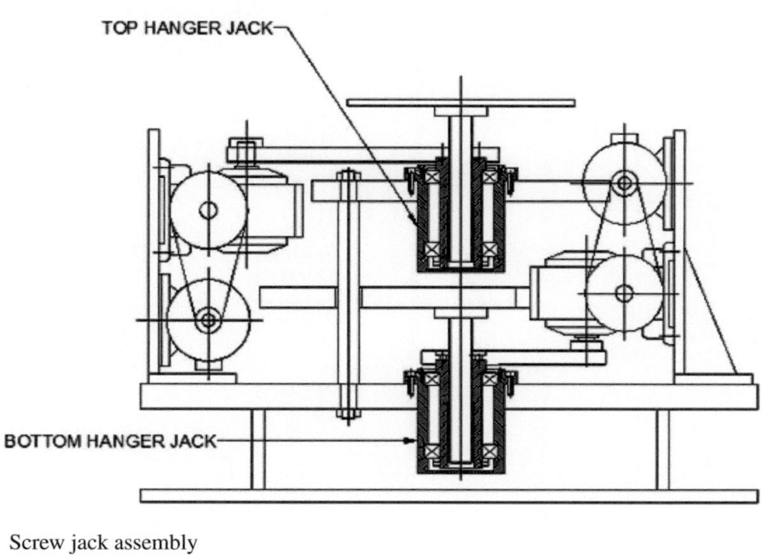

Fig. 2 Screw jack assembly

weight hanger. These motors are programmed to handle the loading and unloading of weights. At the same time, the current setup employs machined square-threaded screws for the jack mechanism.

2.2 *Working of Software*

The software is developed in Java programming language. The algorithm for working the machine is shown in Fig. 3. The software is developed by referring to this algorithm.

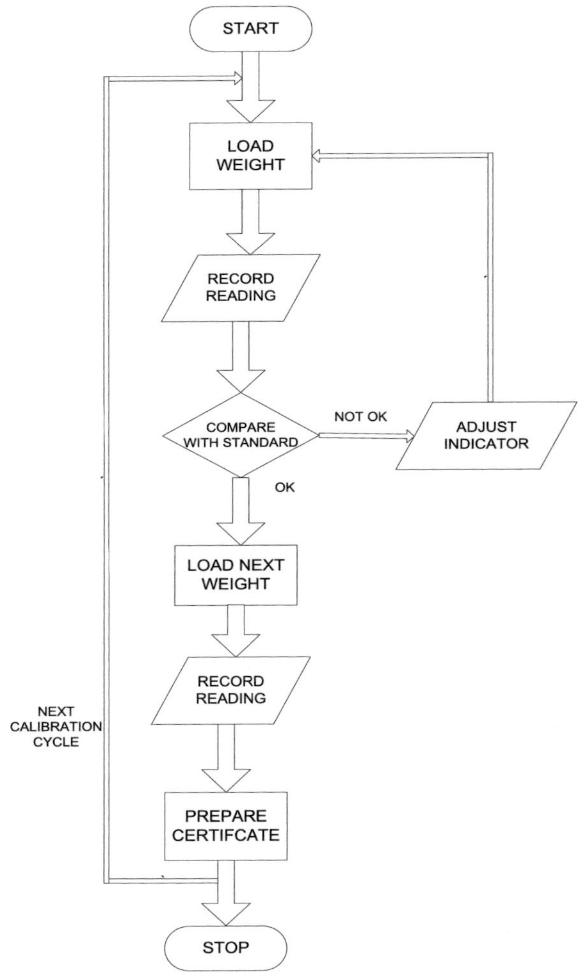

Fig. 3 Algorithm of machine operation

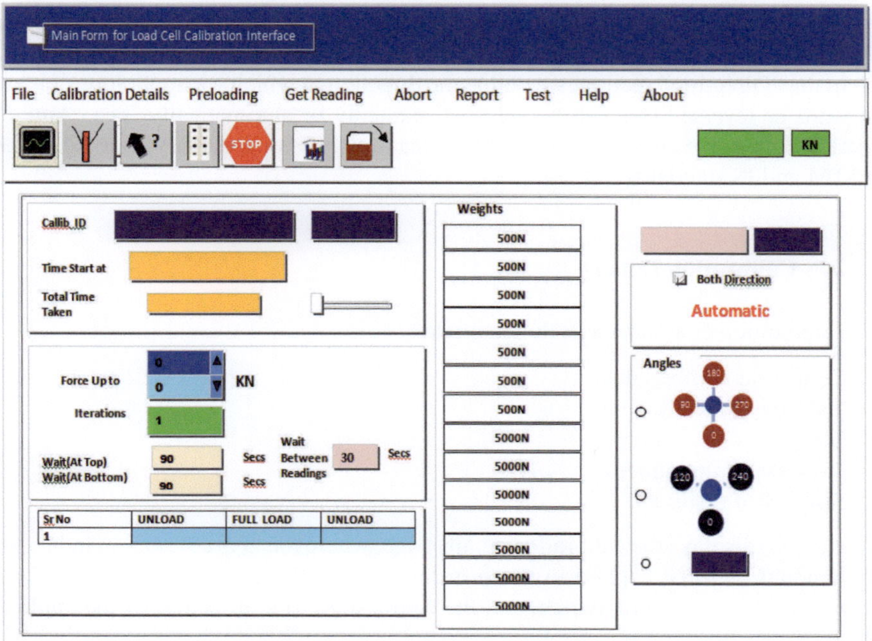

Fig. 4 Main page

The software is developed in easy formats to enter inputs. The flexible structure of the software helps the operator with the error-free operation of the machine.

Figure 4 shows the main form of the force calibration interface in which there are seven sections along with "Help" and "About." The description of each section is as follows:

2.2.1 File

The Institute generates a unique "Calibration ID" for each calibration. In the "Time Starts at" and "Total Time Taken" sections, we record the time taken for each calibration. By analyzing this data, we can define procedures to reduce the time cycle if possible.

In the "Force Up to" section, we determine the force on the calibration instrument as specified by the customer in the service request form. Based on this, the number of readings is decided. The "Iterations" section defines the number of reading iterations taken for calibration, per ASTM and IS standards.

The "Wait at Top, Bottom, and in Between" section defines the waiting time for applying force to the instrument under calibration, measured in seconds. In the "Weights" section, we apply force to the calibration instrument using nine 500N weights, which equals 5 KN, and nine 5 KN weights, which in total equals 50 KN. We can select the weight applied to the instrument based on its capacity.

In the "Direction" section, we define the direction in which we apply force, either forward or reverse. This can be selected manually or automatically. The "Angle" section defines the angle at which we take reading iterations, either 90° or 120°, per ASTM and IS standards.

2.2.2 Calibration Details

The calibration contains the following fields:

- Certificate number, date.
- Instrument details.
- Direction of calibration.
- Location of instrument.
- Number of readings.

Figure 5 displays the calibration details page, where the "Force" section indicates the amount of force applied to the instrument. The "Home L2" and "Home L1" represent the home limit switch, and the arrow shows the direction of the applied force.

There are four buttons provided on this page:

– "Read Save" button: By clicking this button, we can read and save the details.

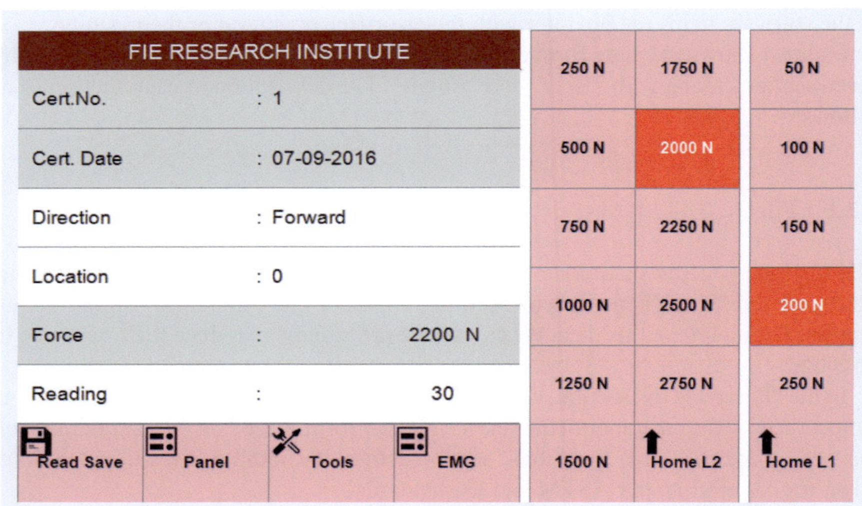

Fig. 5 Calibration details

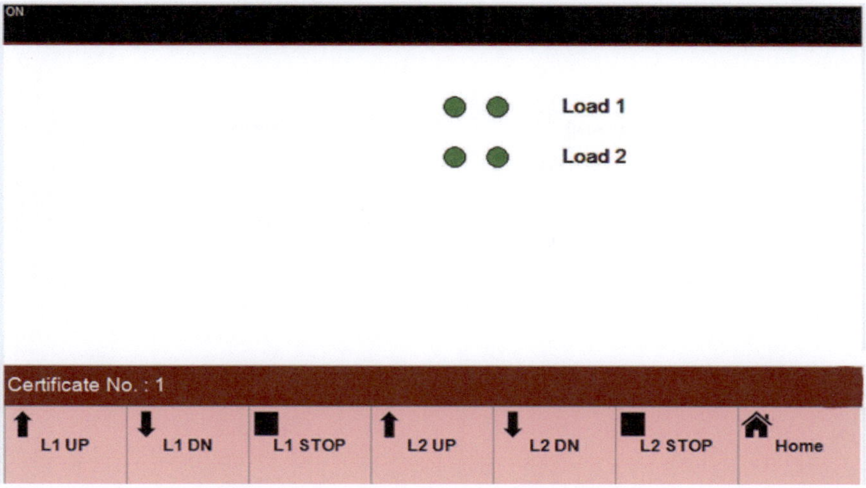

Fig. 6 Pushbutton

- "Panel" section: This section displays a push-button panel that is a replica of the physical push-button panel.
- "Tools" section: This section provides some tools for the operator.
- "EMG" button: This is an emergency button that is used when the machine behaves abnormally.

2.2.3 Preloading

In this, we can define the instrument's capacity by determining the number of iterations of the full load.

2.2.4 Get Readings

In this section, we can get and save the readings.

Figure 6 depicts a Push Button assembly that consists of Load1 weighing 500 N and Load2 weighing 5 KN. L1 UP and DN are used to move Load1 up and down, while L2 UP and DN are used to move Load2 up and down. L1 STOP and L2 STOP can be used to stop the movement of Load1 and Load2, respectively.

2.2.5 Report

Figure 7 shows the report page in this we can "Export" the report in another format like Excel, PDF, etc.

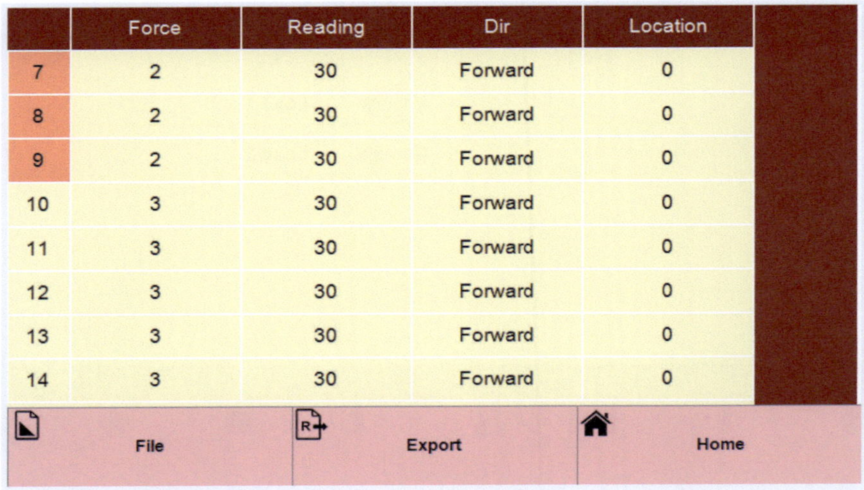

	Force	Reading	Dir	Location
7	2	30	Forward	0
8	2	30	Forward	0
9	2	30	Forward	0
10	3	30	Forward	0
11	3	30	Forward	0
12	3	30	Forward	0
13	3	30	Forward	0
14	3	30	Forward	0

Fig. 7 Report 1

Figure 8 displays the section of the report that contains all the customer details. Additionally, you will see several buttons in this section, each serving a unique purpose as explained below.

Open This option allows you to access existing files.
New You can create a new file by selecting this button.
Save Choose this option to save the details.
Online This button establishes a communication connection between the PC software and the hardware on the machine.
Data This option is relevant to the database.

2.3 Testing and Validation of Machine

The force calibration machines are tested with the help of load cells. The load cells are first tested at the National Metrological Institute, India. The load cells are then checked on the machine, whose accuracy is to be calculated. If the machine shows the readings within tolerance, then it is considered an accurate machine.

2.3.1 Calculating the Dead Weight Force Uncertainty

The vertical force that a stationary mass in a dead weight force machine in air exerts on its support is expressed in SI units as follows:

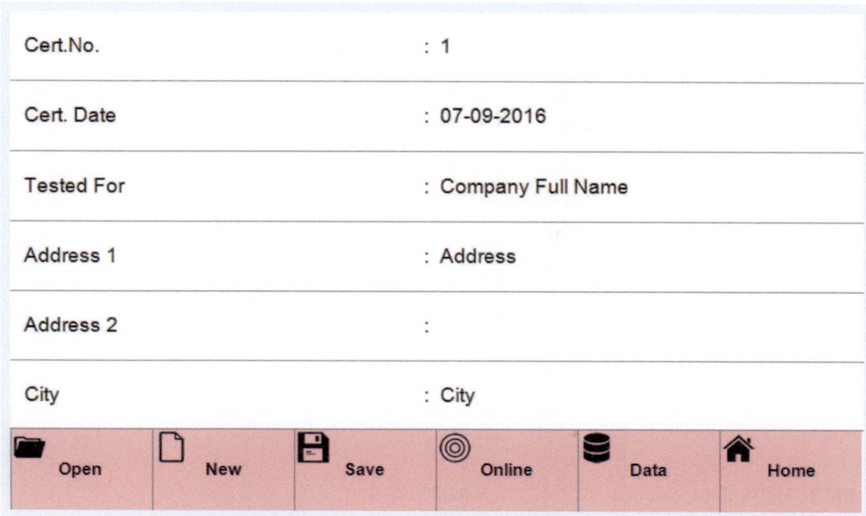

Fig. 8 Report 2

$$F = \left(g - g'\right) m \left(1 - a'/m'\right) \quad (1)$$

where m′ is the mass in kilograms (kg), g′ is the local gravity that is measured close to the machine's bottom, g′ is the variation of g along the machine's height, and a′ and m′ are the densities of the material and air, respectively [10, 11].

Equation (1) makes it clear that the uncertainty of each of the individual parameters of the equation determines the total uncertainty in the force realized by the dead weight force machine at a specific position. Since there is less than a ppm difference in g from the bottom to the top of the loading stack, the term _g in the above equation can be ignored. Based on the additional variations of Eq. (1), the estimated error for the dead weight force calibration machine is ±0.00055 percent (k = 2), as indicated in Tables 1 and 2.

The interaction between the machine and the load cell, the force parasitic components of the force calibrating machine, the measurement process with specific reference to creep and creep recovery, etc. would all prevent this theoretically estimated value from being true in practice and add a finite amount to the estimated total uncertainty associated with the force realized by the machine. Considering the various factors that influence input quantities, discussion, and recommendation, it is assumed that the expanded uncertainty of the vertical components of force applied throughout the entire range of the dead weight force calibrating machine is less than ± 0.005% on a conservative basis [1, 12].

Table 1 Uncertainty calculations for 20 KN load cell

F	X1	X2	X3	X4	X	Xr	Xwr	b
2	0.19990	0.19991	0.19991	0.19990	0.19990	0.19990	0.200	0.0050
4	0.39980	0.39976	0.39974	0.39977	0.39977	0.39977	0.400	0.0150
6	0.59973	0.59975	0.59976	0.59976	0.59975	0.59975	0.600	0.0050
8	0.79961	0.79958	0.79960	0.79961	0.79961	0.79961	0.800	0.0013
10	0.99934	0.99929	0.99930	0.99932	0.99932	0.99932	0.999	0.0040
12	1.19904	1.19908	1.19907	1.19906	1.19906	1.19906	1.199	0.0025
14	1.39865	1.39864	1.39866	1.39865	1.39865	1.39865	1.399	0.0007
16	1.59843	1.59846	1.59845	1.59846	1.59845	1.59845	1.598	0.0019
18	1.79800	1.79801	1.79803	1.79801	1.79801	1.79801	1.798	0.0017
20	1.99751	1.99753	1.99754	1.99753	1.99753	1.99753	1.998	0.0015

Table 2 Uncertainty calculations for 20 KN load cell

b'	fc	Urpd	Urpt	Ures	Ucreep	Uzero	Uint	Ucal	Uc	Uexp
0.0050	0.0101	0.0017	0.0029	1.4E-03	0.0000	0.0000	0.0041	0.0115	0.013	0.025
−0.0100	−0.0158	0.0043	−0.0058	7.2E-04	0.0000	0.0000	−0.0064	0.0115	0.015	0.030
0.0033	0.0046	0.0017	0.0019	4.8E-04	0.0000	0.0000	0.0019	0.0115	0.012	0.024
−0.0038	0.0063	0.0004	−0.0022	3.6E-04	0.0000	0.0000	0.0026	0.0115	0.012	0.024
−0.0050	−0.0017	0.0012	−0.0029	2.9E-04	0.0000	0.0000	−0.0007	0.0115	0.012	0.024
0.0033	−0.0010	0.0007	0.0019	2.4E-04	0.0000	0.0000	−0.0004	0.0115	0.012	0.023
−0.0007	−0.0073	0.0002	−0.0004	2.1E-04	0.0000	0.0000	−0.0030	0.0115	0.012	0.024
0.0019	0.0028	0.0006	0.0011	1.8E-04	0.0000	0.0000	0.0011	0.0115	0.012	0.023
0.0006	0.0001	0.0005	0.0003	1.6E-04	0.0000	0.0000	0.0001	0.0115	0.012	0.023
0.0010	−0.0030	0.0004	0.0006	1.4E-04	0.0000	0.0000	−0.0012	0.0115	0.012	0.023

2.3.2 Procedure for Determining the Force-Proving Instrument's Uncertainty

The uncertainty in the force value derived from the interpolation equation, for increasing forces only, at any deflection is known as the calibration uncertainty. The values taken during the calibration are used to compute a combined standard uncertainty w_c, which is given in force units, at each calibration force F.

Equation 2 illustrates how the values collected throughout the calibration, together with other parameters, are used to calculate a combined relative standard uncertainty, or w_c, at each calibration force [9, 13].

$$w_c = \sqrt{\sum_{i=1}^{8} w_i^2} \tag{2}$$

w_1 the relative standard uncertainty related to the force used for calibration.
w_2 the relative standard uncertainty related to the repeatability of the calibration outcomes.
w_3 relative standard uncertainty related to calibration findings repeatability.
w_4 relative standard uncertainty related to indicator resolution.
w_5 relative standard uncertainty due to instrument creep.
w_6 relative standard uncertainty related to output drifting to zero.
w_7 relative standard uncertainty related to the instrument's temperature.
w_8 the interpolation's relative standard uncertainty.

Plotting these combined standard uncertainties versus force yields a least square fit that accounts for all values. An enlarged uncertainty value is obtained by multiplying the coefficients of this fit by the coverage factor, k = 2 [9, 14].

$$W = k \times w_c \tag{3}$$

2.4 Assessment of Dead Weight Force Performance

To reaffirm the uncertainty associated with the force realized by the machine, two precision load cells 20 KN (HBM make, Serial no. 222530053) and 50 KN (HBM make, Serial no. G 71266) were calibrated on the machine under suitable environmental conditions following the calibration procedure NPL-02 C based on ISO 376-2004 and IS 4169:1998. The 20 KN and 50 KN load cells were also calibrated directly against the force standard machine of NMI under similar conditions following the same procedure. To minimize the uncertainty associated with the indicating instrument, a digital indicator (Model AB22A, HBM, Germany, Serial no. 801122765) having a resolution of 0.00001 mV/V was connected to a PC.

Before initiating the calibration process to achieve improved temperature stability, the load cell that was selected for calibration was mounted on the machine and the indicator was turned on for a long enough period. The manufacturers' self-aligning compression pads were used to transfer the load. To reduce the amount of uncertainty in the mean results caused by the loading process and the load cell's creep pattern, a consistent time interval was kept when taking measurements at each force increment [2, 12].

In every calibration, three initial loads at the 0° location were made to the machine's maximum capacity or the load cell, whichever was smaller, and were maintained for 90 s before being reset to zero. The load cell was preloaded only once when it was moved to new locations at 120° and 240°. Every step is taken to reduce the impact of load time. When the force series returned to zero after 90 s, the process was begun again and repeated at each new load cell position to record the observations in either ascending or descending order or in both directions, depending on the situation. All the calibrations for each of the six sets of force measurements were completed for all load cells [5, 14].

By applying two series of forces in ascending sequence at force values of 2, 4, 6, 8, 10, 12, 14, 16, 18, and 20 KN, the 20 KN load cell was calibrated at the 0º position. Additionally, the load cell was calibrated at 120º and 240º positions using two series of forces, or one series in increasing order from 2 to 20 KN.

Likewise, observations were conducted using the same calibration process at 5, 10, 15, 20, 25, 30, 35, 40, 45, and 50 KN distinct force stages in the other instance with a 50 KN cell. To fully calibrate each load cell, observations were made.

2.5 Calculations of Uncertainty for 20 KN Load Cell

The load cell readings were acquired as previously described, and the corresponding uncertainty calculations have been presented in Tables 1 and 2. The force values are displayed in the first column, followed by four columns (X1, X2, X3, X4) containing readings at 0°, 120°, and 270° rotations. The "X" column represents the average of the readings, while "Xr" and "Xwr" denote the averages with and without rotation, respectively. The "b" and "b'" columns signify reproducibility error and repeatability error, and "Fc" represents interpolation error, calculated through a polynomial equation based on the readings.

Subsequently, columns "Urpd," "Urpt," "Ures," "Ucreep," "Uzero," "Uint," and "Ucal" depict uncertainties related to reproducibility, repeatability, resolution, creep, zero drift, interpolation, and calibration force. The combined standard uncertainty (Uc) is then computed using Eq. 1 with the uncertainties. The expanded uncertainty (Uexp) is determined by multiplying the combined uncertainty by the coverage factor ($k = 2$). All these calculations were done by referring to ISO 376 standards. This comprehensive process of uncertainty calculations was executed for the 20 KN load cell at FRI, and a similar methodology was applied to the 50 KN load cell. Comparative results are elucidated in the subsequent section [9–11].

Tables 3 and 4 list the repeatability, reproducibility, and uncertainty data that were found during the calibration of 20 KN and 50 KN load cells. The increased relative errors seen in the tests made with a 20 KN load cell that was below 20% of its capacity, however, might be caused by the bottom end of the load cell's unique behavior.

The repeatability and reproducibility of the calibrated load cells on this machine are displayed in Figs. (9–14). When these precision load cells are calibrated directly against the dead weight force machine in the optimal range (20% to 100%) of their capacities, these results demonstrate the dependability and efficiency of the created machine. By comparing the repeatability and reproducibility of the observation made at NMI using the same load cells in comparable conditions and after the same calibration process, this may be assessed.

Table 3 Reproducibility, repeatability, and uncertainty of 20 KN load cell

20 KN load cell						
Reproducibility			Repeatability		Uncertainty	
Load	NMI	FRI	NMI	FRI	NMI	FRI
2	0.0350	0.0050	0.0050	0.0050	0.022	0.025
4	0.0225	0.0150	0.0050	−0.0100	0.016	0.030
6	0.0250	0.0050	0.0017	0.0033	0.016	0.024
8	0.0188	0.0013	−0.0013	−0.0038	0.013	0.024
10	0.0080	0.0040	0.0030	−0.0050	0.009	0.024
12	0.0075	0.0025	0.0025	0.0033	0.009	0.023
14	0.0072	0.0007	0.0000	−0.0007	0.008	0.024
16	0.0063	0.0019	0.0013	0.0019	0.008	0.023
18	0.0072	0.0017	−0.0006	0.0006	0.008	0.023
20	0.0050	0.0015	0.0000	0.0010	0.008	0.023

Table 4 Reproducibility, repeatability, and uncertainty of 50 KN load cell

50 KN load cell						
Reproducibility			Repeatability		Uncertainty	
Load	NMI	FRI	NMI	FRI	NMI	FRI
5	0.0490	0.0147	0.0000	0.0245	0.029	0.043
10	0.0343	0.0074	−0.0049	−0.0074	0.022	0.029
15	0.0311	0.0082	0.0000	0.0082	0.019	0.025
20	0.0257	0.0074	−0.0012	0.0074	0.017	0.025
25	0.0206	0.0039	0.0000	0.0039	0.014	0.025
30	0.0188	0.0033	−0.0008	0.0016	0.014	0.025
35	0.0140	0.0028	0.0007	−0.0014	0.012	0.023
40	0.0129	0.0012	0.0006	−0.0018	0.013	0.023
45	0.0114	0.0022	0.0005	0.0022	0.014	0.024
50	0.0098	0.0015	0.0015	0.0015	0.015	0.025

3 Discussion

The automation process for the 50 KN deadweight force calibration machine was successfully developed. After the automation was implemented, it was observed that the calibration process became easier and faster than the manual mode. The entire process took almost 30% less time. Furthermore, the reproducibility, repeatability, and uncertainty aspects of the calibration process are discussed below.

Comparison of Reproducibility

The graphs shown in Figs. 9 and 10 represent reproducibility at FRI and NMI. The reproducibility for the 20 KN load cell is found within the range of 0.0007 to 0.0150, whereas for the 50 KN load cell it falls between 0.0012 and 0.0147. The variation in reproducibility with respective to load can be observed in the graph. For lower loads, reproducibility is comparatively lower than that of at higher loads. The reproducibility of the machine at FRI seems good and compatible with NMIs. For 50KN load, reproducibility variation is quite uniform than for 20 KN load cell.

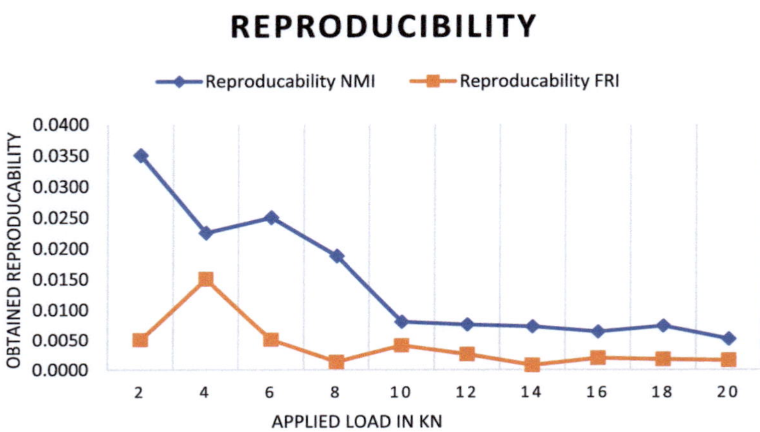

Fig. 9 Reproducibility of 20 KN load cell (NMI versus FRI)

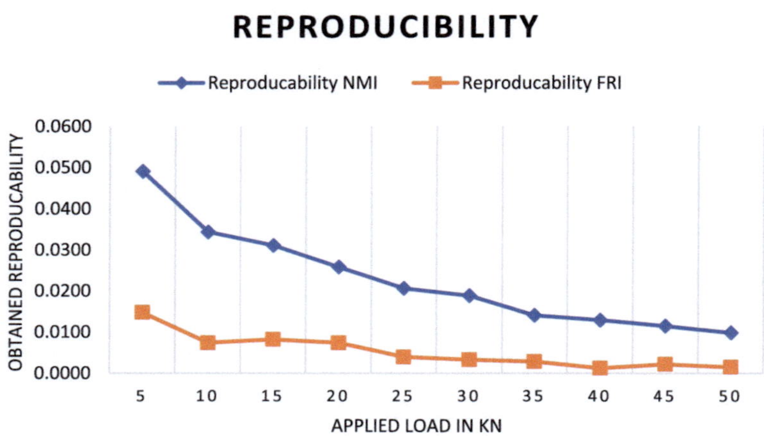

Fig. 10 Reproducibility of 50 KN load cell (NMI versus FRI)

Comparison of Repeatability

The repeatability of the machine with 20 KN and 50 KN load cells is shown in Figs. 11 and 12. The repeatability at NMI is also shown for comparison. The repeatability for the 20 KN load cell is found within the range of −0.010 to 0.0050, whereas for the 50 KN load cell it falls between −0.0074 and 0.0245. It can be observed from the graph that for lower loads repeatability is less as compared with higher loads. The variation for 4 to 10 KN was found more than others. For highest loads, i.e., 20 KN and 50 KN, the repeatability seems almost the same as NMIs.

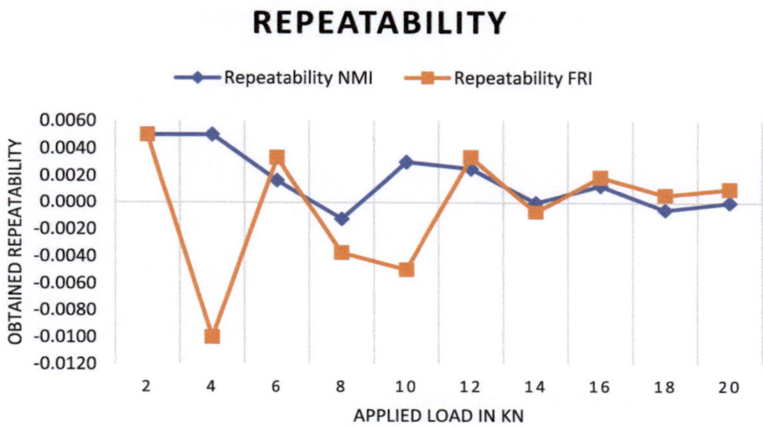

Fig. 11 Repeatability of 20 KN load cell (NMI versus FRI)

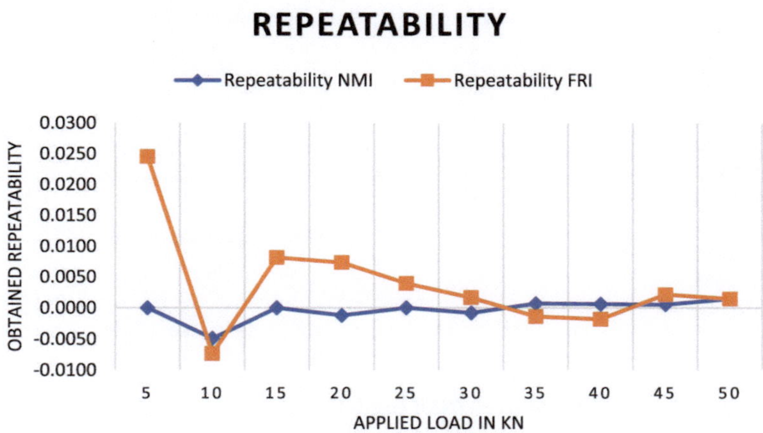

Fig. 12 Repeatability of 50 KN load cell (NMI versus FRI)

Comparison of Uncertainty

The uncertainty of the machine with 20 KN and 50 KN load cells is shown in Figs. 13 and 14. The uncertainty at NMI is also shown for comparison. The uncertainty for the 20 KN load cell is found within the range of 0.023 to 0.030, whereas for the 50 KN load cell it falls between −0.023 and 0.043. It can be observed from the graph that for lower loads uncertainty is more as compared with higher loads.

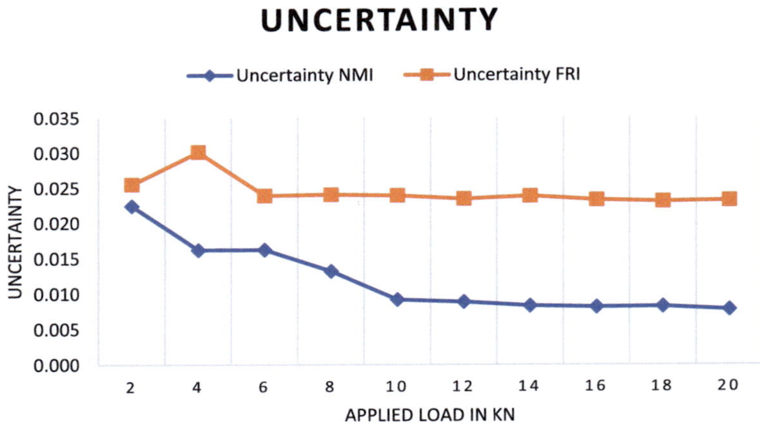

Fig. 13 Uncertainty of 20 KN load cell (NMI versus FRI)

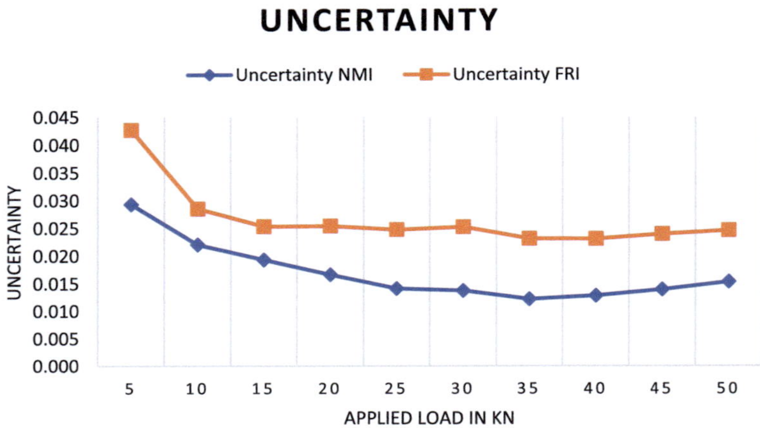

Fig. 14 Uncertainty of 50 KN load cell (NMI versus FRI)

4 Conclusion

To calibrate load cells, a fully automated 50 KN dead weight force machine has been constructed and put into use. High precision is ensured by this system's adherence to globally approved calibration processes. There is a noticeable convergence in the repeatability and reproducibility data when direct calibration of 20 KN and 50 KN load cells is compared. Regarding stability and parasitic components, the system shows a consistent uncertainty of $\pm$ 0.005% over the full range of applied forces.

Through uncertainty estimates, relative reproducibility error, and relative repeatability error, the research of the two precision force transducers (20 KN and 50 KN) is summarized. For both the manual and automatic modes, the computed results for repeatability, reproducibility, and uncertainty are visually displayed. Notably, for both force transducers, the automatic mode exhibits better metrological performance. This study confirms that force transducers can be calibrated efficiently using an automated calibration system in conjunction with the automation of a 50 KN dead weight force machine.

The benefits of automation are shown by the metrological characterization of force transducers with the 50 KN dead weight force machine. Better performance makes it possible to measure intricate sequences; loading time intervals may be precisely controlled; and reliable indicator readings reduce operator dependence. The machine readings' close alignment with the NMI readings confirms that the automation can successfully calibrate load cells within the required range.

To sum up, the automated method shows notable improvements in accuracy, repeatability, and reproducibility, proving to be a dependable load cell calibration tool.

References

1. Kumar H (2013) A contemporary approach for evaluation of the best measurement capability of a force calibration machine. Int J Mod Phys Conf Ser 24(Apmf):1360013. https://doi.org/10.1142/s2010194513600136
2. Xu C (1998) A practical model for uncertainty evaluation in force measurements. Meas Sci Technol 9(11):1831–1836. https://doi.org/10.1088/0957-0233/9/11/006
3. Singh AP, Ghoshal SK, Kumar H (2020) Development and metrological evaluation of an industrial force transducer. Mapan J Metrol Soc India 35(2):165–176. https://doi.org/10.1007/s12647-019-00359-z
4. Force calibration_what is it_how does it work_types & process
5. Liang W, Yang X, Yao J, Wei T (2020) Calibration of force standard machine by build-up force transfer system in the highest range up to 60 MN. Mapan J Metrol Soc India 35(2):151–164. https://doi.org/10.1007/s12647-019-00362-4
6. Company MI (2022) Conditions, methods, and systems that impact force calibration, pp 1–96
7. Company MI (2022) Force calibration guidance for beginners
8. Mitchell RA et al (1986) National Bureau of Standards force calibration at the National Bureau of Standards
9. Euromet (2011) Uncertainty of force measurements (March)

10. Maybank L, Knott A, Elkington D (1999) Guide to the uncertainty of force measurement (April)
11. Knott A et al (2022) EURAMET calibration guide no. 4—guidelines on the uncertainty of force measurements
12. Titus SSK, Dhulkhe SK, Yadav P, Jain KK (2009) Development and performance evaluation of a dead weight force machine in 2–50 N range. Mapan J Metrol Soc India 24(4):225–232. https://doi.org/10.1007/s12647-009-0027-5
13. Company MI (2021) ISO 376 guidance and uncertainty measurements relating to force equipment, pp 1–28
14. Kumar H, Kumar A, Yadav P (2011) Improved performance of 50 kN dead weight force machine using automation as a tool. Meas Sci Rev 11(2):67–70. https://doi.org/10.2478/v10048-011-0013-2

FSC
www.fsc.org
MIX
Papier aus verantwortungsvollen Quellen
Paper from responsible sources
FSC® C105338

If you have any concerns about our products,
you can contact us on
ProductSafety@springernature.com

In case Publisher is established outside the EU,
the EU authorized representative is:
Springer Nature Customer Service Center GmbH
Europaplatz 3, 69115 Heidelberg, Germany

Printed by Libri Plureos GmbH
in Hamburg, Germany